ESSENTIALS OF BOTANY

Edited by

Tadena, Anjanette S.

Arcler Press

Essentials of Botany

Edited by Tadena, Anjanette S.

Edition 2018

ArclerPress

Published by
Arcler Press LLC
708 3rd Avenue, 6th Floor
New York
NY 10017
United States of America
www.arclerpress.com

ISBN: 978-1-68094-543-0 (Hardback)

Printed in United States of America

Library of Congress Control Number: 2017933113

About the Editor

Tadena, Anjanette S.

Anjanette obtained her degree from the University of the Philippines. She is also working on her Master's Degree in Environmental Science Specializing in Environmental Toxicity. Her fascination outside work are astrophotography and traveling. She is currently working as a Research Fellow at the Bureau of Agriculture and Fisheries Standards - Department of Agriculture, Philippines.

Table of Contents

3 PLANT MORPHOLOGY

4 FLOWER REPRODUCTION

5 PLANT PHYSIOLOGY

6 PLANT TAXONOMY

7 PLANT ANATOMY

8 PLANT ECOLOGY

9 CONTROL OF PLANT DISEASES

List of Abbreviations

ABA	Abscissic Acid
ATP	Adenosine Triphosphate
AP	APETALA,
AGP	Arabinogalactan Proteins
AAPFCO	Association of American Plant Food Control Officials
BES	British Ecological Society
Cl	Chlorine
DNA	Deoxyribonucleic Acid
FBP	Floral Binding Protein
GP	Glyceraldehyde -phosphate
GRPs	Glycine-rich proteins
HRGP	Hydroxyproline-rich Glycoproteins
IAA	Indole--acetic Acid
LFY	LEAFY,
MFs	Microfilaments,
MTs	Microtubules,
PGRs	Plant growth Regulators
RNA	Ribonucleic Acid

Preface

Botany is the scientific study of plants. Plants, means a wide range of living organisms from the smallest bacteria to the largest living things - the giant sequoia trees. Whether you call it botany, plant science or plant biology, it is one of the world's oldest natural sciences. By this definition plants include: algae, fungi, lichens, mosses, ferns, conifers and flowering plants. Today scientists believe bacteria, algae and fungi are in their own distinct kingdoms, but most general botany courses, and most Botany Departments at colleges and universities, still teach about these groups. Because the field is so broad, there are many kinds of plant biologists and many different opportunities available. Botanists interested in ecology study interactions of plants with other organisms and the environment. Other field botanists search to find new species or do experiments to discover how plants grow under different conditions. Some botanists study the structure of plants. They may work in the field, concentrating on the pattern of the whole plant. Others use microscopes to study the most detailed fine structure of individual cells. Many botanists do experiments to determine how plants convert simple chemical compounds into more complex chemicals. Plants are vital to all life on earth. They convert light from the sun into energy we can eat, and even use for fuel. They also release the oxygen we breathe. Botanists study plants to try to understand how these processes work. Their research is involved in enhancing crops, developing medicines, cleaning up contaminated sites, and even powering our cars. It's an exciting field on the cutting edge of the clean energy economy. It's also a profession for nature lovers. Botanists study processes that occur on a time scale ranging from fractions of a second in individual cells to those that unfold over eons of evolutionary time.

Content Coverage

Chapter One introduces the basic concepts of botany such as history and scope as well as its importance.

Chapters Two presents an emphasis on Plants group, are multicellular eukaryotes of the kingdom Plantae. They form an unranked clade Viridiplantae that includes the flowering plants, conifers and other gymnosperms, ferns, clubmosses, hornworts, liverworts, mosses and the green algae.

Chapters Three and *Four* discuss about the plant morphology and reproduction in flowering plants. Reproduction in flowering plants begins with pollination, the transfer of pollen from anther to stigma on the same flower or to the stigma of another flower on the same plant (self-pollination), or from anther on one plant to the stigma of another plant (cross-pollination).

Chapter Five presents a comprehensive study of all the internal activities of plants—those chemical and physical processes associated with life as they occur in plants.

Chapters Six and *Seven* are plant taxonomy and plant anatomy, separately.

Chapters Eight and *Nine* discuss about plant ecology and plant pathology, respectively. Plant ecology studies the distribution and abundance of plants, the effects of environmental factors upon the abundance of plants, and the interactions among and between plants and other organisms. Control of plant diseases is crucial to the reliable production of food, and it provides significant reductions in agricultural use of land, water, fuel and other inputs.

CHAPTER 1

Botany: An Introduction

Chapter Contents

- **Introduction**
- **History**
- **Scope and Importance**
- **Plant Biochemistry**
- **Genetics**
- **Plant Evolution**
- **Plant Physiology**
- **Plant Anatomy and Morphology**
- **Systematic Botany**

Botany is the scientific study of plants. "Plants," to most people, means a wide range of living organisms from the smallest bacteria to the largest living things - the giant sequoia trees. By this definition plants include: algae, fungi, lichens, mosses, ferns, conifers and flowering plants. Today scientists believe bacteria, algae and fungi are in their own distinct kingdoms, but most general botany courses, and most Botany Departments at colleges and universities, still teach about these groups.

INTRODUCTION

Botany, also called plant science(s), plant biology or phytology, is the science of plant life and a branch of biology. A botanist or plant scientist is a scientist who specialises in this field. The term "botany" comes from the Ancient Greek word βοτάνη (botanē) meaning "pasture", "grass", or "fodder"; βοτάνη is in turn derived from βόσκειν (boskein), "to feed" or "to graze". Traditionally, botany has also included the study of fungi and algae by mycologists and phycologists respectively, with the study of these three groups of organisms remaining within the sphere of interest of the International Botanical Congress. Nowadays, botanists study approximately 400,000 species of living organisms of which some 260,000 species are vascular plants and about 248,000 are flowering plants.

The fruit of Myristica fragrans, a species native to Indonesia, is the source of two valuable spices, the red aril (mace) enclosing the dark brown nutmeg.

Botany originated in prehistory as herbalism with the efforts of early humans to identify – and later cultivate – edible, medicinal and poisonous plants, making it one of the oldest branches of science. Medieval physic gardens, often attached to monasteries, contained plants of medical importance. They were forerunners of the first botanical gardens attached to universities, founded from the 1540s onwards. One of the earliest was the Padua botanical garden. These gardens facilitated the academic study of plants. Efforts to catalogue and describe their collections were the beginnings of plant taxonomy, and led in 1753 to the binomial system of Carl Linnaeus that remains in use to this day.

In the 19th and 20th centuries, new techniques were developed for the study of plants, including methods of optical microscopy and live cell imaging, electron microscopy, analysis of chromosome number, plant chemistry and the structure and function of enzymes and other proteins. In the last two decades of the 20th century, botanists exploited the techniques of molecular genetic analysis, including genomics and proteomics and DNA sequences to classify plants more accurately.

Modern botany is a broad, multidisciplinary subject with inputs from most other areas of science and technology. Research topics include the study of plant structure, growth and differentiation, reproduction, biochemistry and primary metabolism, chemical products, development, diseases, evolutionary relationships, systematics, and plant taxonomy. Dominant themes in 21st century plant science are molecular genetics and epigenetics, which are the mechanisms and control of gene expression during differentiation of plant cells and tissues. Botanical research has diverse applications in providing staple foods, materials such as timber, oil, rubber, fibre and drugs, in modern horticulture, agriculture and forestry, plant propagation, breeding and genetic modification, in the synthesis of chemicals and raw materials for construction and energy production, in environmental management, and the maintenance of biodiversity.

HISTORY

Early Botany

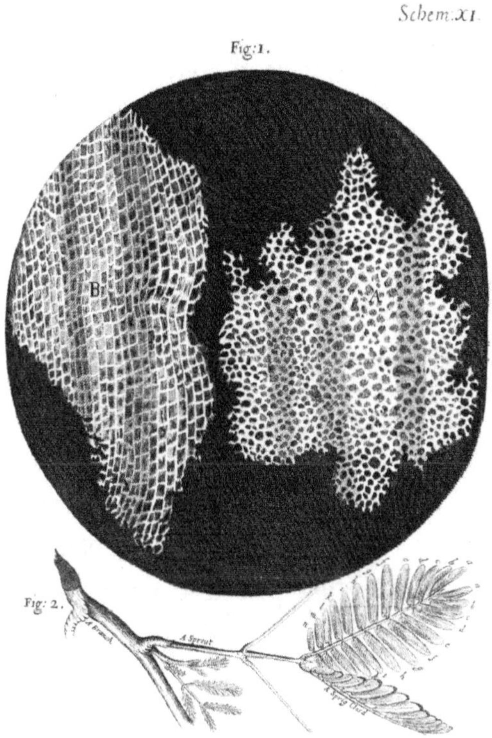

An engraving of the cells of cork, from Robert Hooke's Micrographia, 1665

Botany originated as herbalism, the study and use of plants for their medicinal properties. Many records of the Holocene period date early botanical knowledge as far back as 10,000 years ago. This early unrecorded knowledge of plants was discovered in ancient sites of human occupation within Tennessee, which make up much of the Cherokee land today. The early recorded history of botany includes many ancient writings and plant classifications. Examples of early botanical works have been found in ancient texts from India dating back to before 1100 BC, in archaic Avestan writings, and in works from China before it was unified in 221 BC.

Modern botany traces its roots back to Ancient Greece specifically to Theophrastus (c. 371–287 BC), a student of Aristotle who invented and described many of its principles and is widely regarded in the scientific community as the "Father of Botany". His major works, Enquiry into Plants and On the Causes of Plants, constitute the most important contributions to botanical science until the Middle Ages, almost seventeen centuries later. Another work from Ancient Greece that made an early impact on botany is De Materia Medica, a five-volume encyclopedia about herbal medicine written in the middle of the first century by Greek physician and pharmacologist Pedanius Dioscorides. De Materia Medica was widely read for more than 1,500 years. Important contributions from the medieval Muslim world include Ibn Wahshiyya's Nabatean Agriculture, Abū anīfa Dīnawarī's (828–896) the Book of Plants, and Ibn Bassal's The Classification of Soils. In the early 13th century, Abu al-Abbas al-Nabati, and Ibn al-Baitar (d. 1248) wrote on botany in a systematic and scientific manner.

In the mid-16th century, "botanical gardens" were founded in a number of Italian universities – the Padua botanical garden in 1545 is usually considered to be the first which is still in its original location. These gardens continued the practical value of earlier "physic gardens", often associated with monasteries, in which plants were cultivated for medical use. They supported the growth of botany as an academic subject. Lectures were given about the plants grown in the gardens and their medical uses demonstrated. Botanical gardens came much later to northern Europe; the first in England was the University of Oxford Botanic Garden in 1621. Throughout this period, botany remained firmly subordinate to medicine.

German physician Leonhart Fuchs (1501–1566) was one of "the three German fathers of botany", along with theologian Otto Brunfels (1489–1534) and physician Hieronymus Bock (1498–1554) (also called Hieronymus Tragus). Fuchs and Brunfels broke away from the tradition of copying earlier works to make original observations of their own. Bock created his own system of plant classification.

Physician Valerius Cordus (1515–1544) authored a botanically and pharmacologically important herbal Historia Plantarum in 1544 and a pharmacopoeia of lasting importance, the Dispensatorium in 1546. Naturalist Conrad von Gesner (1516–1565) and herbalist John Gerard (1545–c. 1611) published herbals covering the medicinal uses of plants. Naturalist Ulisse Aldrovandi (1522–1605) was considered the father of natural history, which included the study of plants. In 1665, using an early microscope, Polymath Robert Hooke discovered cells, a term he coined, in cork, and a short time later in living plant tissue.

Early Modern Botany

The Linnaean Garden of Linnaeus' residence in Uppsala, Sweden, was planted according to his Systema sexuale.

During the 18th century, systems of plant identification were developed comparable to dichotomous keys, where unidentified plants are placed into taxonomic groups (e.g. family, genus and species) by making a series of choices between pairs of characters. The choice and sequence of the characters may be artificial in keys designed purely for identification (diagnostic keys) or more closely related to the natural or phyletic order of the taxa in synoptic keys. By the 18th century, new plants for study were arriving in Europe in increasing numbers from newly discovered countries and the European colonies worldwide. In 1753 Carl von Linné (Carl Linnaeus) published his Species Plantarum, a hierarchical classification of plant species that remains the reference point for modern botanical nomenclature. This established a standardised binomial or two-part naming scheme where the first name represented the genus and the second identified the species within the genus. For the purposes of identification, Linnaeus's Systema Sexuale classified plants into 24 groups according to the number of their male sexual organs. The 24th group, Cryptogamia, included all plants with concealed reproductive parts, mosses, liverworts, ferns, algae and fungi.

Increasing knowledge of plant anatomy, morphology and life cycles led to the realisation that there were more natural affinities between plants than the artificial sexual system of Linnaeus. Adanson (1763), de Jussieu (1789), and Candolle (1819) all proposed various alternative natural systems of classification that grouped plants using a wider range of shared characters and were widely followed. The Candollean system reflected his ideas of the progression of morphological complexity and the later classification by Bentham and Hooker, which was influential until the mid-19th century, was influenced by Candolle's approach. Darwin's publication of the Origin of Species in 1859 and his concept of common descent required modifications to the Candollean system to reflect evolutionary relationships as distinct from mere morphological similarity.

Botany was greatly stimulated by the appearance of the first "modern" textbook, Matthias Schleiden's Grundzüge der Wissenschaftlichen Botanik, published in English in 1849 as Principles of Scientific Botany. Schleiden was a microscopist and an early plant anatomist who co-founded the cell theory with Theodor Schwann and Rudolf Virchow and was among the first to grasp the significance of the cell nucleus that had been described by Robert Brown in 1831. In 1855, Adolf Fick formulated Fick's laws that enabled the calculation of the rates of molecular diffusion in biological systems.

Echeveria glauca in a Connecticut greenhouse. Botany uses Latin names for identification, here, the specific name glauca means blue.

Modern Botany

Micropropagation of transgenic plants

Biologist and statistician Ronald Fisher

Building upon the gene-chromosome theory of heredity that originated with Gregor Mendel (1822–1884), August Weismann (1834–1914) proved that inheritance only takes place through gametes. No other cells can pass on inherited characters.The work of

Katherine Esau (1898–1997) on plant anatomy is still a major foundation of modern botany. Her books Plant Anatomy and Anatomy of Seed Plants have been key plant structural biology texts for more than half a century.

The discipline of plant ecology was pioneered in the late 19th century by botanists such as Eugenius Warming, who produced the hypothesis that plants form communities, and his mentor and successor Christen C. Raunkiær whose system for describing plant life forms is still in use today. The concept that the composition of plant communities such as temperate broadleaf forest changes by a process of ecological succession was developed by Henry Chandler Cowles, Arthur Tansley and Frederic Clements. Clements is credited with the idea of climax vegetation as the most complex vegetation that an environment can support and Tansley introduced the concept of ecosystems to biology. Building on the extensive earlier work of Alphonse de Candolle, Nikolai Vavilov (1887–1943) produced accounts of the biogeography, centres of origin, and evolutionary history of economic plants.

Particularly since the mid-1960s there have been advances in understanding of the physics of plant physiological processes such as transpiration (the transport of water within plant tissues), the temperature dependence of rates of water evaporation from the leaf surface and the molecular diffusion of water vapour and carbon dioxide through stomatal apertures. These developments, coupled with new methods for measuring the size of stomatal apertures, and the rate of photosynthesis have enabled precise description of the rates of gas exchange between plants and the atmosphere. Innovations in statistical analysis by Ronald Fisher, Frank Yates and others at Rothamsted Experimental Station facilitated rational experimental design and data analysis in botanical research. The discovery and identification of the auxin plant hormones by Kenneth V. Thimann in 1948 enabled regulation of plant growth by externally applied chemicals. Frederick Campion Steward pioneered techniques of micropropagation and plant tissue culture controlled by plant hormones. The synthetic auxin 2,4-Dichlorophenoxyacetic acid or 2,4-D was one of the first commercial synthetic herbicides.

20th century developments in plant biochemistry have been driven by modern techniques of organic chemical analysis, such as spectroscopy, chromatography and electrophoresis. With the rise of the related molecular-scale biological approaches of molecular biology, genomics, proteomics and metabolomics, the relationship between the plant genome and most aspects of the biochemistry, physiology, morphology and behaviour of plants can be subjected to detailed experimental analysis. The concept originally stated by Gottlieb Haberlandt in 1902 that all plant cells are totipotent and can be grown in vitro ultimately enabled the use of genetic engineering experimentally to knock out a gene or genes responsible for a specific trait, or to add genes such as GFP that report when a gene of interest is being expressed. These technologies enable the biotechnological use of whole plants or plant cell cultures grown in bioreactors to synthesise pesticides, antibiotics or other pharmaceuticals, as well as the practical application of genetically modified crops designed for traits such as improved yield.

Modern morphology recognises a continuum between the major morphological categories of root, stem (caulome), leaf (phyllome) and trichome. Furthermore, it emphasises structural dynamics. Modern systematics aims to reflect and discover phylogenetic relationships between plants. Modern Molecular phylogenetics largely ignores morphological characters, relying on DNA sequences as data. Molecular analysis of DNA sequences from most families of flowering plants enabled the Angiosperm Phylogeny Group to publish in 1998 a phylogeny of flowering plants, answering many of the questions about relationships among angiosperm families and species. The theoretical possibility of a practical method for identification of plant species and commercial varieties by DNA barcoding is the subject of active current research.

SCOPE AND IMPORTANCE

Botany involves the recording and description of plants, such as this herbarium specimen of the lady fern Athyrium filix-femina.

The study of plants is vital because they underpin almost all animal life on Earth by generating a large proportion of the oxygen and food that provide humans and other organisms with aerobic respiration with the chemical energy they need to exist. Plants, algae and cyanobacteria are the major groups of organisms that carry out photosynthesis, a process that uses the energy of sunlight to convert water and carbon dioxide into sugars that can be used both as a source of chemical energy and of organic molecules that are used in the structural components of cells. As a by-product of photosynthesis, plants release oxygen into the atmosphere, a gas that is required by nearly all living things to carry out cellular respiration. In addition, they are influential in the global carbon and water cycles and plant roots bind and stabilise soils, preventing soil erosion. Plants are crucial to the future of human society as they provide food, oxygen, medicine, and products for people, as well as creating and preserving soil. Historically, all living things were classified as either animals or plants and botany covered the study of all organisms not considered animals. Botanists examine both the internal functions and processes within plant organelles, cells, tissues, whole plants, plant populations and plant communities. At each of these levels, a botanist may be concerned with the classification (taxonomy), phylogeny and evolution, structure (anatomy and morphology), or function (physiology) of plant life.

The strictest definition of "plant" includes only the "land plants" or embryophytes, which include seed plants (gymnosperms, including the pines, and flowering plants) and the free-sporing cryptogams including ferns, clubmosses, liverworts, hornworts and mosses. Embryophytes are multicellular eukaryotes descended from an ancestor that obtained its energy from sunlight by photosynthesis. They have life cycles with alternating haploid and diploid phases. The sexual haploid phase of embryophytes, known as the gametophyte, nurtures the developing diploid embryo sporophyte within its tissues for at least part of its life, even in the seed plants, where the gametophyte itself is nurtured by its parent sporophyte. Other groups of organisms that were previously studied by botanists include bacteria (now studied in bacteriology), fungi (mycology) – including lichen-forming fungi (lichenology), non-chlorophyte algae (phycology), and viruses (virology). However, attention is still given to these groups by botanists, and fungi (including lichens) and photosynthetic protists are usually covered in introductory botany courses.

Palaeobotanists study ancient plants in the fossil record to provide information about the evolutionary history of plants. Cyanobacteria, the first oxygen-releasing photosynthetic organisms on Earth, are thought to have given rise to the ancestor of plants by entering into an endosymbiotic relationship with an early eukaryote, ultimately becoming the chloroplasts in plant cells. The new photosynthetic plants (along with their algal relatives) accelerated the rise in atmospheric oxygen started by the cyanobacteria, changing the ancient oxygen-free, reducing, atmosphere to one in which free oxygen has been abundant for more than 2 billion years.

Among the important botanical questions of the 21st century are the role of plants as primary producers in the global cycling of life's basic ingredients: energy, carbon, oxygen, nitrogen and water, and ways that our plant stewardship can help address the global environmental issues of resource management, conservation, human food security, biologically invasive organisms, carbon sequestration, climate change, and sustainability.

Human Nutrition

The food we eat comes directly or indirectly from plants such as rice.

Virtually all staple foods come either directly from primary production by plants, or indirectly from animals that eat them.Plants and other photosynthetic organisms are at the base of most food chains because they use the energy from the sun and nutrients from the soil and atmosphere, converting them into a form that can be used by animals. This is what ecologists call the first trophic level. The modern forms of the major staple foods, such as maize, rice, wheat and other cereal grasses, pulses, bananas and plantains, as well as flax and cotton grown for their fibres, are the outcome of prehistoric selection over thousands of years from among wild ancestral plants with the most desirable characteristics.

Botanists study how plants produce food and how to increase yields, for example through plant breeding, making their work important to mankind's ability to feed the world and provide food security for future generations. Botanists also study weeds, which are a considerable problem in agriculture, and the biology and control of plant pathogens in agriculture and natural ecosystems. Ethnobotany is the study of the relationships between plants and people. When applied to the investigation of historical plant–people relationships ethnobotany may be referred to as archaeobotany or palaeoethnobotany. Some of the earliest plant-people relationships arose between the indigenous people of Canada in identifying edible plants from inedible plants. This relationship the indigenous people had with plants was recorded by ethnobotanists.

PLANT BIOCHEMISTRY

Plant biochemistry is the study of the chemical processes used by plants. Some of these processes are used in their primary metabolism like the photosynthetic Calvin cycle and crassulacean acid metabolism. Others make specialised materials like the cellulose and lignin used to build their bodies, and secondary products like resins and aroma compounds.

Plants make various photosynthetic pigments, some of which can be seen here through paper chromatography.

Plants and various other groups of photosynthetic eukaryotes collectively known as "algae" have unique organelles known as chloroplasts. Chloroplasts are thought to be descended from cyanobacteria that formed endosymbiotic relationships with ancient plant and algal ancestors. Chloroplasts and cyanobacteria contain the blue-green pigment chlorophyll a. Chlorophyll a (as well as its plant and green algal-specific cousin chlorophyll b)[a] absorbs light in the blue-violet and orange/red parts of the spectrum while reflecting and transmitting the green light that we see as the characteristic colour of these organisms. The energy in the red and blue light that these pigments absorb is used by chloroplasts to make energy-rich carbon compounds from carbon dioxide and water by oxygenic photosynthesis, a process that generates molecular oxygen (O_2) as a by-product.

The Calvin cycle (Interactive diagram) The Calvin cycle incorporates carbon dioxide into sugar molecules.

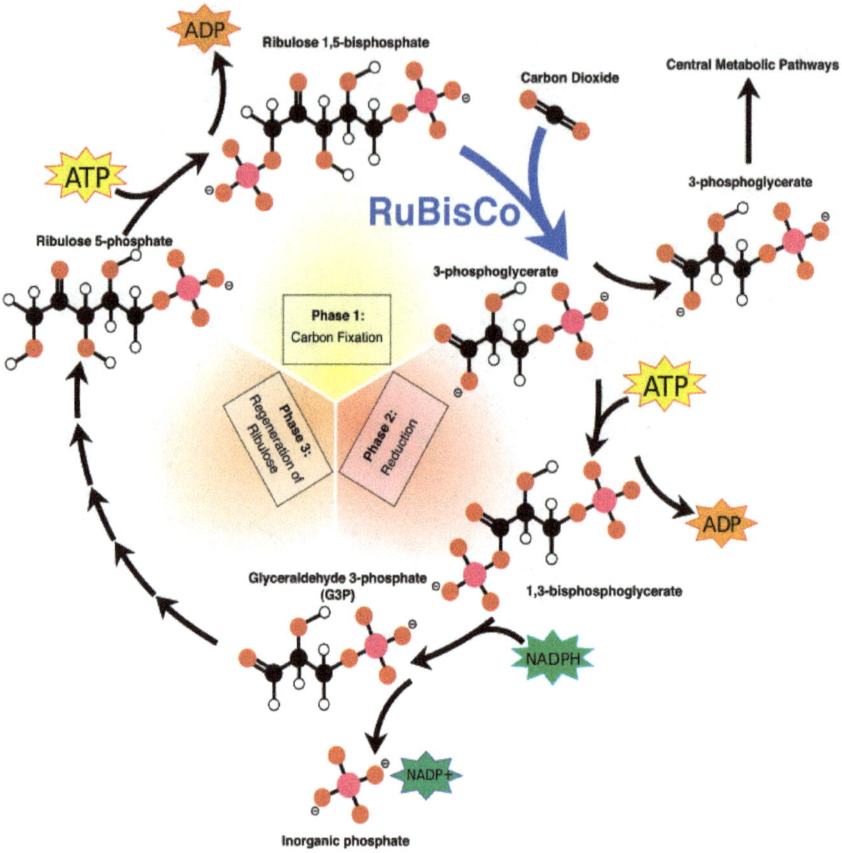
ADP
Ribulose 1,5-bisphosphate
Carbon Dioxide
Central Metabolic Pathways
ATP
RuBisCo
3-phosphoglycerate
Ribulose 5-phosphate
3-phosphoglycerate
Phase 1:
Carbon Fixation
ATP
Phase 3:
Regeneration of
Ribulose
Phase 2:
Reduction
ADP
Glyceraldehyde 3-phosphate
(G3P)
1,3-bisphosphoglycerate
NADPH
NADP+
Inorganic phosphate

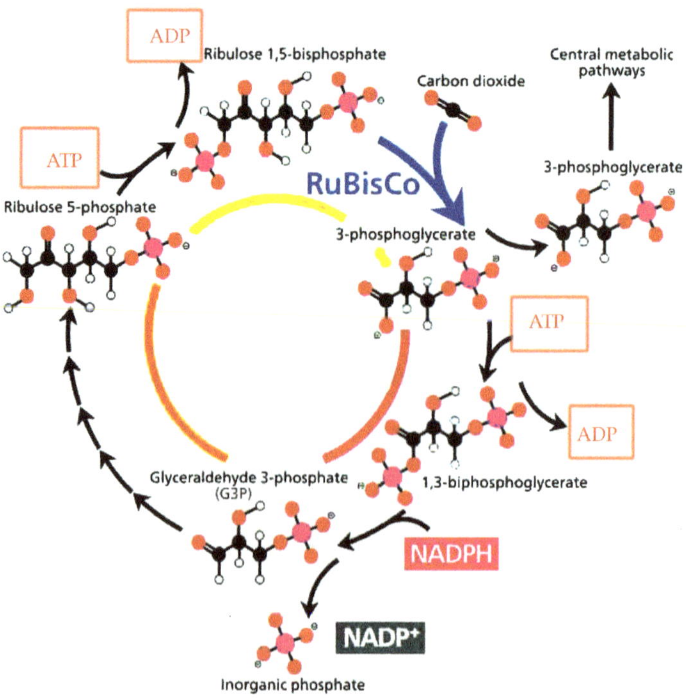
ADP
Ribulose 1,5-bisphosphate
Carbon dioxide
Central metabolic pathways
ATP
RuBisCo
3-phosphoglycerate
Ribulose 5-phosphate
3-phosphoglycerate
ATP
ADP
Glyceraldehyde 3-phosphate
(G3P)
1,3-biphosphoglycerate
NADPH
NADP+
Inorganic phosphate

The light energy captured by chlorophyll a is initially in the form of electrons (and later a proton gradient) that's used to make molecules of ATP and NADPH which temporarily store and transport energy. Their energy is used in the light-independent reactions of the Calvin cycle by the enzyme rubisco to produce molecules of the 3-carbon sugar glyceraldehyde 3-phosphate (G3P). Glyceraldehyde 3-phosphate is the first product of photosynthesis and the raw material from which glucose and almost all other organic molecules of biological origin are synthesised. Some of the glucose is converted to starch which is stored in the chloroplast. Starch is the characteristic energy store of most land plants and algae, while inulin, a polymer of fructose is used for the same purpose in the sunflower family Asteraceae. Some of the glucose is converted to sucrose (common table sugar) for export to the rest of the plant.

Unlike in animals (which lack chloroplasts), plants and their eukaryote relatives have delegated many biochemical roles to their chloroplasts, including synthesising all their fatty acids, and most amino acids. The fatty acids that chloroplasts make are used for many things, such as providing material to build cell membranes out of and making the polymer cutin which is found in the plant cuticle that protects land plants from drying out.

Plants synthesise a number of unique polymers like the polysaccharide molecules cellulose, pectin and xyloglucan from which the land plant cell wall is constructed. Vascular land plants make lignin, a polymer used to strengthen the secondary cell walls of xylem tracheids and vessels to keep them from collapsing when a plant sucks water through them under water stress. Lignin is also used in other cell types like sclerenchyma fibres that provide structural support for a plant and is a major constituent of wood. Sporopollenin is a chemically resistant polymer found in the outer cell walls of spores and pollen of land plants responsible for the survival of early land plant spores and the pollen of seed plants in the fossil record. It is widely regarded as a marker for the start of land plant evolution during the Ordovician period. The concentration of carbon dioxide in the atmosphere today is much lower than it was when plants emerged onto land during the Ordovician and Silurian periods. Many monocots like maize and the pineapple and some dicots like the Asteraceae have since independently evolved pathways like Crassulacean acid metabolism and the C_4 carbon fixation pathway for photosynthesis which avoid the losses resulting from photorespiration in the more common C_3 carbon fixation pathway. These biochemical strategies are unique to land plants.

Medicine and Materials

Phytochemistry is a branch of plant biochemistry primarily concerned with the chemical substances produced by plants during secondary metabolism. Some of these compounds are toxins such as the alkaloid coniine from hemlock. Others, such as the essential oils peppermint oil and lemon oil are useful for their aroma, as flavourings and spices (e.g., capsaicin), and in medicine as pharmaceuticals as in opium from opium poppies. Many medicinal and recreational drugs, such as tetrahydrocannabinol (active ingredient in cannabis), caffeine, morphine and nicotine come directly from plants. Others are simple derivatives of botanical natural products. For example, the pain killer aspirin is the acetyl ester of salicylic acid, originally isolated from the bark of willow trees, and a wide range of opiate painkillers like heroin are obtained by chemical modification of morphine obtained from the opium poppy. Popular stimulants come from plants, such as caffeine from coffee, tea and chocolate, and nicotine from tobacco. Most alcoholic beverages come from fermentation of carbohydrate-rich plant products such as barley (beer), rice (sake) and grapes (wine). Native Americans have used various plants as ways of treating illness or disease for thousands of years. This knowledge Native Americans have on plants has been recorded by enthnobotanists and then in turn has been used by pharmaceutical companies as a way of drug discovery.

Tapping a rubber tree in Thailand

Plants can synthesise useful coloured dyes and pigments such as the anthocyanins responsible for the red colour of red wine, yellow weld and blue woad used together to produce Lincoln green, indoxyl, source of the blue dye indigo traditionally used to dye denim and the artist's pigments gamboge and rose madder.

Sugar, starch, cotton, linen, hemp, some types of rope, wood and particle boards, papyrus and paper, vegetable oils, wax, and natural rubber are examples of commercially important materials made from plant tissues or their secondary products. Charcoal, a pure form of carbon made by pyrolysis of wood, has a long history as a metal-smelting fuel, as a filter material and adsorbent and as an artist's material and is one of the three ingredients of gunpowder. Cellulose, the world's most abundant organic polymer, can be converted into energy, fuels, materials and chemical feedstock. Products made from cellulose include rayon and cellophane, wallpaper paste, biobutanol and gun cotton. Sugarcane, rapeseed and soy are some of the plants with a highly fermentable sugar or oil content that are used as sources of biofuels, important alternatives to fossil fuels, such as biodiesel. Sweetgrass was used by NativeAmericanse to ward of bugs like mosquitoes. These bug repelling properties of sweetgrass were later found by the American Chemical Society in the molecules phytol and coumarin.

Plant ecology

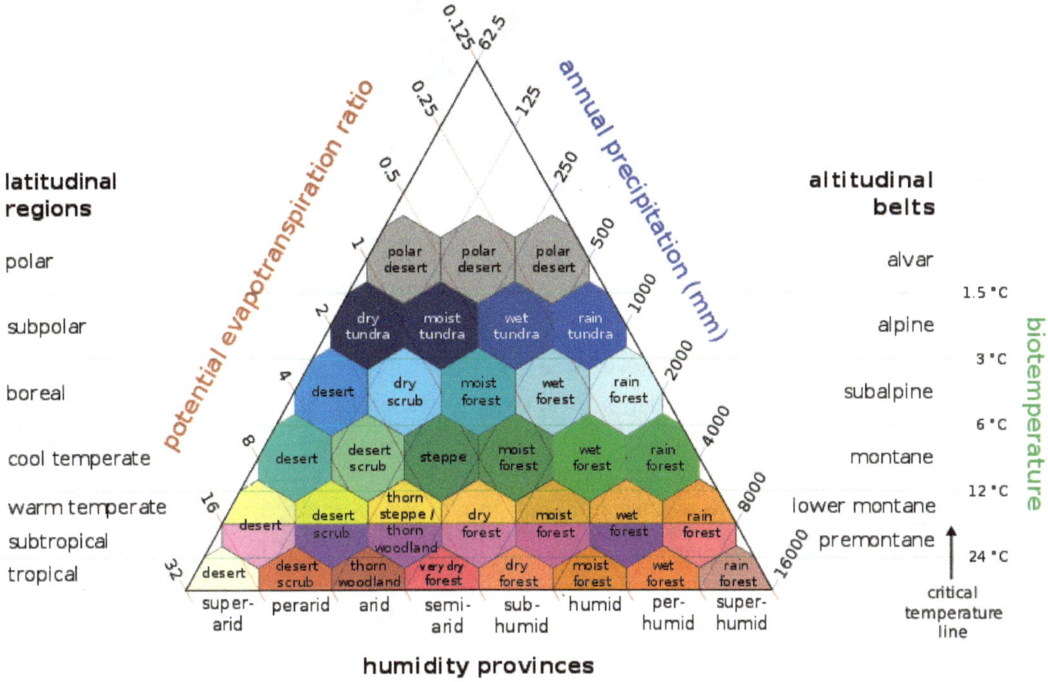

Holdridge life zones model relationships between vegetation type, moisture availability and temperature.

Plant ecology is the science of the functional relationships between plants and their habitats—the environments where they complete their life cycles. Plant ecologists study the composition of local and regional floras, their biodiversity, genetic diversity and fitness, the adaptation of plants to their environment, and their competitive or mutualistic interactions with other species. Some ecologists even rely on empirical data from indigenous people that is gathered by ethnobotanists. This information can relay a great deal of information on how the land once was thousands of years ago and how it has changed over that time. The goals of plant ecology are to understand the causes of their distribution patterns, productivity, environmental impact, evolution, and responses to environmental change.

Plants depend on certain edaphic (soil) and climatic factors in their environment but can modify these factors too. For example, they can change their environment's albedo, increase runoff interception, stabilise mineral soils and develop their organic content, and affect local temperature. Plants compete with other organisms in their ecosystem for resources. They interact with their neighbours at a variety of spatial scales in groups, populations and communities that collectively constitute vegetation. Regions with characteristic vegetation types and dominant plants as well as similar abiotic and biotic factors, climate, and geography make up biomes like tundra or tropical rainforest.

The nodules of Medicago italica contain the nitrogen fixing bacterium Sinorhizobium meliloti. The plant provides the bacteria with nutrients and an anaerobic environment, and the bacteria fix nitrogen for the plant.

Herbivores eat plants, but plants can defend themselves and some species are parasitic or even carnivorous. Other organisms form mutually beneficial relationships with plants. For example, mycorrhizal fungi and rhizobia provide plants with nutrients in exchange for food, ants are recruited by ant plants to provide protection, honey bees, bats and other animals pollinate flowers and humans and other animals act as dispersal vectors to spread spores and seeds.

Plants, Climate and Environmental Change

Plant responses to climate and other environmental changes can inform our understanding of how these changes affect ecosystem function and productivity. For example, plant phenology can be a useful proxy for temperature in historical climatology, and the biological impact of climate change and global warming. Palynology, the analysis of fossil pollen deposits in sediments from thousands or millions of years ago allows the reconstruction of past climates. Estimates of atmospheric CO_2 concentrations since the Palaeozoic have been obtained from stomatal densities and the leaf shapes and sizes of ancient land plants. Ozone depletion can expose plants to higher levels of ultraviolet radiation-B (UV-B), resulting in lower growth rates. Moreover, information from studies of community ecology, plant systematics, and taxonomy is essential to understanding vegetation change, habitat destruction and species extinction.

GENETICS

Inheritance in plants follows the same fundamental principles of genetics as in other multicellular organisms. Gregor Mendel discovered the genetic laws of inheritance by studying inherited traits such as shape in Pisum sativum (peas). What Mendel learned from studying plants has had far reaching benefits outside of botany. Similarly, "jumping genes" were discovered by Barbara McClintock while she was studying maize. Nevertheless, there are some distinctive genetic differences between plants and other organisms.

Species boundaries in plants may be weaker than in animals, and cross species hybrids are often possible. A familiar example is peppermint, Mentha × piperita, a sterile hybrid between Mentha aquatica and spearmint, Mentha spicata.The many cultivated varieties of wheat are the result of multiple inter- and intra-specific crosses between wild species and their hybrids. Angiosperms with monoecious flowers often have self-incompatibility mechanisms that operate between the pollen and stigma so that the pollen either fails to reach the stigma or fails to germinate and produce male gametes. This is one of several methods used by plants to promote outcrossing. In many land plants the male and female gametes are produced by separate individuals. These species are said to be dioecious when referring to vascular plant sporophytes and dioicous when referring to bryophyte gametophytes.

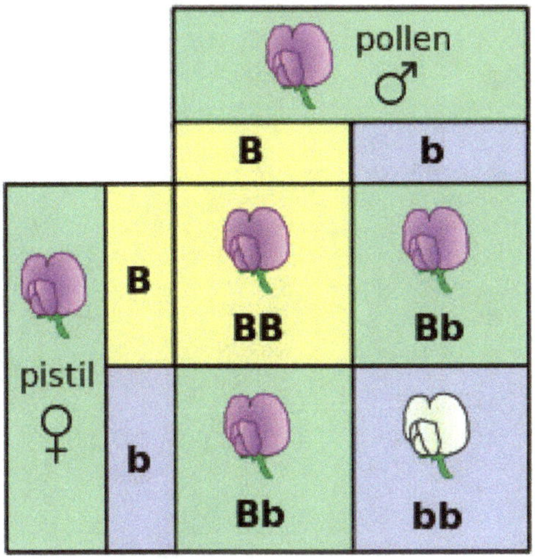

A Punnett square depicting a cross between two pea plants heterozygous for purple (B) and white (b) blossoms

Unlike in higher animals, where parthenogenesis is rare, asexual reproduction may occur in plants by several different mechanisms. The formation of stem tubers in potato is one example. Particularly in arctic or alpine habitats, where opportunities for fertilisation of flowers by animals are rare, plantlets or bulbs, may develop instead of flowers, replacing sexual reproduction with asexual reproduction and giving rise to clonal populations genetically identical to the parent. This is one of several types of apomixis that occur in plants. Apomixis can also happen in a seed, producing a seed that contains an embryo genetically identical to the parent.

Most sexually reproducing organisms are diploid, with paired chromosomes, but doubling of their chromosome number may occur due to errors in cytokinesis. This can occur early in development to produce an autopolyploid or partly autopolyploid organism, or during normal processes of cellular differentiation to produce some cell types that are polyploid (endopolyploidy), or during gamete formation. An allopolyploid plant may result from a hybridisation event between two different species. Both autopolyploid and allopolyploid plants can often reproduce normally, but may be unable to cross-breed successfully with the parent population because there is a mismatch in chromosome numbers. These plants that are reproductively isolated from the parent species but live within the same geographical area, may be sufficiently successful to form a new species. Some otherwise sterile plant polyploids can still reproduce vegetatively or by seed apomixis, forming clonal populations of identical individuals. Durum wheat is a fertile tetraploid allopolyploid, while bread wheat is a fertile hexaploid. The commercial banana is an example of a sterile, seedless triploid hybrid. Common dandelion is a triploid that produces viable seeds by apomictic seed.

As in other eukaryotes, the inheritance of endosymbiotic organelles like mitochondria and chloroplasts in plants is non-Mendelian. Chloroplasts are inherited through the male parent in gymnosperms but often through the female parent in flowering plants.

Molecular Genetics

Thale cress, Arabidopsis thaliana, the first plant to have its genome sequenced, remains the most important model organism.

A considerable amount of new knowledge about plant function comes from studies of the molecular genetics of model plants such as the Thale cress, Arabidopsis thaliana, a weedy species in the mustard family (Brassicaceae). The genome or hereditary information contained in the genes of this species is encoded by about 135 million base pairs of DNA, forming one of the smallest genomes among flowering plants. Arabidopsis was the first plant to have its genome sequenced, in 2000.The sequencing of some other relatively small genomes, of rice (Oryza sativa) and Brachypodium distachyon, has made them important model species for understanding the genetics, cellular and molecular biology of cereals, grasses and monocots generally.

Model plants such as Arabidopsis thaliana are used for studying the molecular biology of plant cells and the chloroplast. Ideally, these organisms have small genomes that are well known or completely sequenced, small stature and short generation times. Corn has been used to study mechanisms of photosynthesis and phloem loading of sugar in C_4 plants.The single celled green alga Chlamydomonas reinhardtii, while not an embryophyte itself, contains a green-pigmentedchloroplast related to that of land plants, making it useful for study. A red alga Cyanidioschyzon merolae has also been used to study some basic chloroplast functions. Spinach, peas, soybeans and a moss Physcomitrella patens are commonly used to study plant cell biology.

Agrobacterium tumefaciens, a soil rhizosphere bacterium, can attach to plant cells and infect them with a callus-inducing Ti plasmid by horizontal gene transfer, causing a callus infection called crown gall disease. Schell and Van Montagu (1977) hypothesised that the Ti plasmid could be a natural vector for introducing the Nif gene responsible for nitrogen fixation in the root nodules of legumes and other plant species. Today, genetic modification of the Ti plasmid is one of the main techniques for introduction of transgenes to plants and the creation of genetically modified crops.

Epigenetics

Epigenetics is the study of heritable changes in gene function that cannot be explained by changes in the underlying DNA sequence but cause the organism's genes to behave (or "express themselves") differently. One example of epigenetic change is the marking of the genes by DNA methylation which determines whether they will be expressed or not. Gene expression can also be controlled by repressor proteins that attach to silencer regions of the DNA and prevent that region of the DNA code from being expressed. Epigenetic marks may be added or removed from the DNA during programmed stages of development of the plant, and are responsible, for example, for the differences between anthers, petals and normal leaves, despite the fact that they all have the same underlying genetic code. Epigenetic changes may be temporary or may remain through successive cell divisions for the remainder of the cell's life. Some epigenetic changes have been shown to be heritable, while others are reset in the germ cells.

Epigenetic changes in eukaryotic biology serve to regulate the process of cellular differentiation. During morphogenesis, totipotent stem cells become the various pluripotent cell lines of the embryo, which in turn become fully differentiated cells. A single fertilised egg cell, the zygote, gives rise to the many different plant cell types including parenchyma, xylem vessel elements, phloem sieve tubes, guard cells of the epidermis, etc. as it continues to divide. The process results from the epigenetic activation of some genes and inhibition of others.

Unlike animals, many plant cells, particularly those of the parenchyma, do not terminally differentiate, remaining totipotent with the ability to give rise to a new individual plant. Exceptions include highly lignified cells, the sclerenchyma and xylem which are dead at maturity, and the phloem sieve tubes which lack nuclei. While plants use many of the same epigenetic mechanisms as animals, such as chromatin remodelling, an alternative hypothesis is that plants set their gene expression patterns using positional information from the environment and surrounding cells to determine their developmental fate.

PLANT EVOLUTION

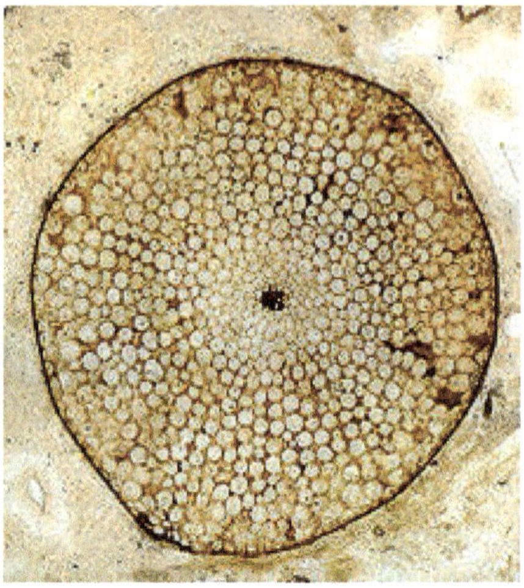

Transverse section of a fossil stem of the Devonian vascular plant Rhynia gwynne-vaughani

The chloroplasts of plants have a number of biochemical, structural and genetic similarities to cyanobacteria, (commonly but incorrectly known as "blue-green algae") and are thought to be derived from an ancient endosymbiotic relationship between an ancestral eukaryotic cell and a cyanobacterial resident.

The algae are a polyphyletic group and are placed in various divisions, some more closely related to plants than others. There are many differences between them in features such as cell wall composition, biochemistry, pigmentation, chloroplast structure and nutrient reserves. The algal division Charophyta, sister to the green algal division Chlorophyta, is considered to contain the ancestor of true plants. The Charophyte class Charophyceae and the land plant sub-kingdom Embryophyta together form the monophyletic group or clade Streptophytina.

Nonvascular land plants are embryophytes that lack the vascular tissues xylem and phloem. They include mosses, liverworts and hornworts. Pteridophytic vascular plants with true xylem and phloem that reproduced by spores germinating into free-living gametophytes evolved during the Silurian period and diversified into several lineages during the late Silurian and early Devonian. Representatives of the lycopods have survived to the present day. By the end of the Devonian period, several groups, including the lycopods, sphenophylls and progymnosperms, had independently evolved "megaspory" – their spores were of two distinct sizes, larger megaspores and smaller microspores. Their reduced gametophytes developed from megaspores retained within the spore-producing organs (megasporangia) of the sporophyte, a condition known as endospory. Seeds consist of an endosporic megasporangium surrounded by one or two sheathing layers (integuments). The young sporophyte develops within the seed, which on germination splits to release it. The earliest known seed plants date from the latest Devonian Famennian stage. Following the evolution of the seed habit, seed plants diversified, giving rise to a number of now-extinct groups, including seed ferns, as well as the modern gymnosperms and angiosperms. Gymnosperms produce "naked seeds" not fully enclosed in an ovary; modern representatives include conifers, cycads, Ginkgo, and Gnetales. Angiosperms produce seeds enclosed in a structure such as a carpel or an ovary. Ongoing research on the molecular phylogenetics of living plants appears to show that the angiosperms are a sister clade to the gymnosperms.

PLANT PHYSIOLOGY

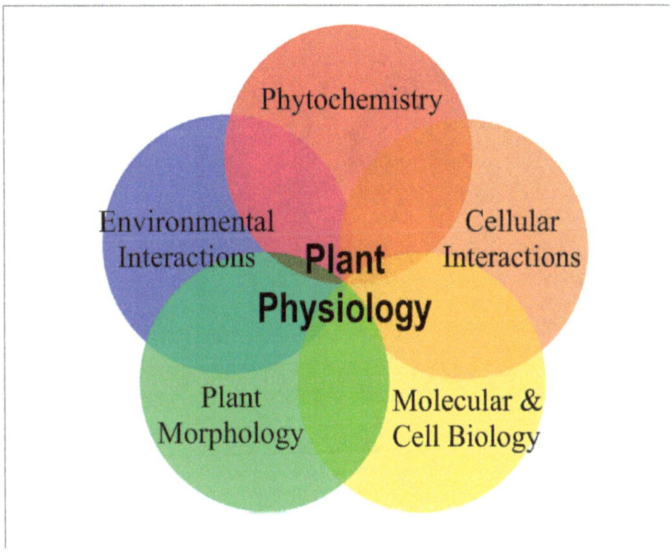

Five of the key areas of study within plant physiology

Plant physiology encompasses all the internal chemical and physical activities of plants associated with life.Chemicals obtained from the air, soil and water form the basis of all plant metabolism. The energy of sunlight, captured by oxygenic photosynthesis and released by cellular respiration, is the basis of almost all life. Photoautotrophs, including all green plants, algae and cyanobacteria gather energy directly from sunlight by photosynthesis. Heterotrophs including all animals, all fungi, all completely parasitic plants, and non-photosynthetic bacteria take in organic molecules produced by photoautotrophs and respire them or use them in the construction of cells and tissues. Respiration is the oxidation of carbon compounds by breaking them down into simpler structures to release the energy they contain, essentially the opposite of photosynthesis.

Molecules are moved within plants by transport processes that operate at a variety of spatial scales. Subcellular transport of ions, electrons and molecules such as water and enzymes occurs across cell membranes. Minerals and water are transported from roots to other parts of the plant in the transpiration stream. Diffusion, osmosis, and active transport and mass flow are all different ways transport can occur. Examples of elements that plants need to transport are nitrogen, phosphorus, potassium, calcium, magnesium, and sulphur. In vascular plants, these elements are extracted from the soil as soluble ions by the roots and transported throughout the plant in the xylem. Most of the elements required for plant nutrition come from the chemical breakdown of soil minerals. Sucrose produced by photosynthesis is transported from the leaves to other parts of the plant in the phloem and plant hormones are transported by a variety of processes.

Plant Hormones

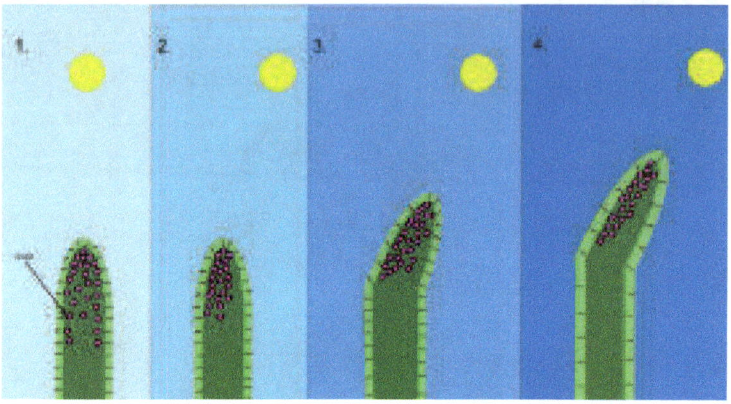

1 An oat coleoptile with the sun overhead. Auxin (pink) is evenly distributed in its tip.

2 With the sun at an angle and only shining on one side of the shoot, auxin moves to the opposite side and stimulates cell elongation there.

3 and **4** Extra growth on that side causes the shoot to bend towards the sun.

Plants are not passive, but respond to external signals such as light, touch, and injury by moving or growing towards or away from the stimulus, as appropriate. Tangible evidence of touch sensitivity is the almost instantaneous collapse of leaflets of Mimosa pudica, the insect traps of Venus flytrap and bladderworts, and the pollinia of orchids.

The hypothesis that plant growth and development is coordinated by plant hormones or plant growth regulators first emerged in the late 19th century. Darwin experimented on the movements of plant shoots and roots towards light and gravity, and concluded "It is hardly an exaggeration to say that the tip of the radicle . . acts like the brain of one of the lower animals . . directing the several movements". About the same time, the role of auxins (from the Greek auxein, to grow) in control of plant growth was first outlined by the Dutch scientist Frits Went.The first known auxin, indole-3-acetic acid (IAA), which promotes cell growth, was only isolated from plants about 50 years later. This compound mediates the tropic responses of shoots and roots towards light and gravity. The finding in 1939 that plant callus could be maintained in culture containing IAA, followed by the observation in 1947 that it could be induced to form roots and shoots by controlling the concentration of growth hormones were key steps in the development of plant biotechnology and genetic modification.

Venus's fly trap, Dionaea muscipula, showing the touch-sensitive insect trap in action

Cytokinins are a class of plant hormones named for their control of cell division or cytokinesis. The natural cytokinin zeatin was discovered in corn, Zea mays, and is a derivative of the purine adenine. Zeatin is produced in roots and transported to shoots in the xylem where it promotes cell division, bud development, and the greening of chloroplasts. The gibberelins, such as Gibberelic acid are diterpenes synthesised from acetyl CoA via the mevalonate pathway. They are involved in the promotion of germination and dormancy-breaking in seeds, in regulation of plant height by controlling stem elongation and the control of flowering. Abscisic acid (ABA) occurs in all land plants except liverworts, and is synthesised from carotenoids in the chloroplasts and other plastids. It inhibits cell division, promotes seed maturation, and dormancy, and promotes stomatal closure. It was so named because it was originally thought to control abscission. Ethylene is a gaseous hormone that is produced in all higher plant tissues from methionine. It is now known to be the hormone that stimulates or regulates fruit ripening and abscission, and it, or the synthetic growth regulator ethephon which is rapidly metabolised to produce ethylene, are used on industrial scale to promote ripening of cotton, pineapples and other climacteric crops.

Another class of phytohormones is the jasmonates, first isolated from the oil of Jasminum grandiflorum which regulates wound responses in plants by unblocking the expression of genes required in the systemic acquired resistance response to pathogen attack.

In addition to being the primary energy source for plants, light functions as a signalling device, providing information to the plant, such as how much sunlight the plant receives each day. This can result in adaptive changes in a process known as photomorphogenesis. Phytochromes are the photoreceptors in a plant that are sensitive to light.

PLANT ANATOMY AND MORPHOLOGY

A nineteenth-century illustration showing the morphology of the roots, stems, leaves and flowers of the rice plant Oryza sativa

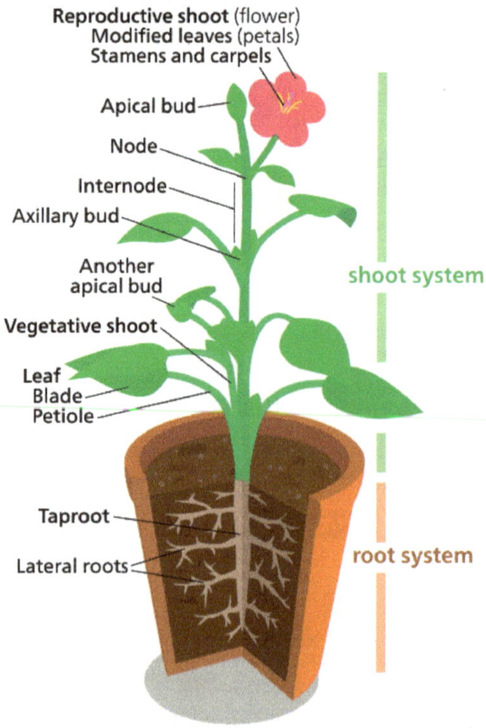

A diagram of a "typical" eudicot, the most common type of plant (three-fifths of all plant species). No plant actually looks exactly like this though.

Plant anatomy is the study of the structure of plant cells and tissues, whereas plant morphology is the study of their external form. All plants are multicellular eukaryotes, their DNA stored in nuclei. The characteristic features of plant cells that distinguish them from those of animals and fungi include a primary cell wall composed of the polysaccharides cellulose, hemicellulose and pectin, larger vacuoles than in animal cells and the presence of plastids with unique photosynthetic and biosynthetic functions as in the chloroplasts. Other plastids contain storage products such as starch (amyloplasts) or lipids (elaioplasts). Uniquely, streptophyte cells and those of the green algal order Trentepohliales divide by construction of a phragmoplast as a template for building a cell plate late in cell division.

The bodies of vascular plants including clubmosses, ferns and seed plants (gymnosperms and angiosperms) generally have aerial and subterranean subsystems. The shoots consist of stems bearing green photosynthesising leaves and reproductive structures. The underground vascularised roots bear root hairs at their tips and generally lack chlorophyll. Non-vascular plants, the liverworts, hornworts and mosses do not produce ground-penetrating vascular roots and most of the plant participates in photosynthesis. The sporophyte generation is nonphotosynthetic in liverworts but may be able to contribute part of its energy needs by photosynthesis in mosses and hornworts.

The root system and the shoot system are interdependent – the usually nonphotosynthetic root system depends on the shoot system for food, and the usually photosynthetic shoot system depends on water and minerals from the root system. Cells in each system are capable of creating cells of the other and producing adventitious shoots or roots. Stolons and tubers are examples of shoots that can grow roots. Roots that spread out close to the surface, such as those of willows, can produce shoots and ultimately new plants. In the event that one of the systems is lost, the other can often regrow it. In fact it is possible to grow an entire plant from a single leaf, as is the case with Saintpaulia, or even a single cell – which can dedifferentiate into a callus (a mass of unspecialised cells) that can grow into a new plant. In vascular plants, the xylem and phloem are the conductive tissues that transport resources between shoots and roots. Roots are often adapted to store food such as sugars or starch, as in sugar beets and carrots.

Stems mainly provide support to the leaves and reproductive structures, but can store water in succulent plants such as cacti, food as in potato tubers, or reproduce vegetatively as in the stolons of strawberry plants or in the process of layering. Leaves gather sunlight and carry out photosynthesis. Large, flat, flexible, green leaves are called foliage leaves. Gymnosperms, such as conifers, cycads, Ginkgo, and gnetophytes are seed-producing plants with open seeds. Angiosperms are seed-producing plants that produce flowers and have enclosed seeds. Woody plants, such as azaleas and oaks, undergo a secondary growth phase resulting in two additional types of tissues: wood (secondary xylem) and bark (secondary phloem and cork). All gymnosperms and many angiosperms are woody plants. Some plants reproduce sexually, some asexually, and some via both means.

Although reference to major morphological categories such as root, stem, leaf, and trichome are useful, one has to keep in mind that these categories are linked through intermediate forms so that a continuum between the categories results. Furthermore, structures can be seen as processes, that is, process combinations.

SYSTEMATIC BOTANY

A botanist preparing a plant specimen for mounting in the herbarium

Systematic botany is part of systematic biology, which is concerned with the range and diversity of organisms and their relationships, particularly as determined by their evolutionary history. It involves, or is related to, biological classification, scientific taxonomy and phylogenetics. Biological classification is the method by which botanists group organisms into categories such as genera or species. Biological classification is a form of scientific taxonomy. Modern taxonomy is rooted in the work of Carl Linnaeus, who grouped species according to shared physical characteristics. These groupings have since been revised to align better with the Darwinian principle of common descent – grouping organisms by ancestry rather than superficial characteristics. While scientists do not always agree on how to classify organisms, molecular phylogenetics, which uses DNA sequences as data, has driven many recent revisions along evolutionary lines and is likely to continue to do so. The dominant classification system is called Linnaean taxonomy. It includes ranks and binomial nomenclature. The nomenclature of botanical organisms is codified in the International Code of Nomenclature for algae, fungi, and plants (ICN) and administered by the International Botanical Congress.

Kingdom Plantae belongs to Domain Eukarya and is broken down recursively until each species is separately classified. The order is: Kingdom; Phylum (or Division); Class; Order; Family; Genus (plural genera); Species. The scientific name of a plant represents its genus and its species within the genus, resulting in a single worldwide name for each organism. For example, the tiger lily is Lilium columbianum. Lilium is the genus, and columbianum the specific epithet. The combination is the name of the species. When writing the scientific name of an organism, it is proper to capitalise the first letter in the genus and put all of the specific epithet in lowercase. Additionally, the entire term is ordinarily italicised (or underlined when italics are not available). The evolutionary relationships and heredity of a group of organisms is called its phylogeny. Phylogenetic studies attempt to discover phylogenies. The basic approach is to use similarities based on shared inheritance to determine relationships. As an example, species of Pereskia are trees or bushes with prominent leaves. They do not obviously resemble a typical leafless cactus such as an Echinocactus. However, both Pereskia and Echinocactus have spines produced from areoles (highly specialised pad-like structures) suggesting that the two genera are indeed related.

Two cacti of Very Different Appearance

Pereskia aculeata

Echinocactus grusonii

Although Pereskia is a tree with leaves, it has spines and areoles like a more typical cactus, such as Echinocactus.

Judging relationships based on shared characters requires care, since plants may resemble one another through convergent evolution in which characters have arisen independently. Some euphorbias have leafless, rounded bodies adapted to water conservation similar to those of globular cacti, but characters such as the structure of their flowers make it clear that the two groups are not closely related. The cladistic method takes a systematic approach to characters, distinguishing between those that carry no information about shared evolutionary history – such as those evolved separately in different groups (homoplasies) or those left over from ancestors (plesiomorphies) – and derived characters, which have been passed down from innovations in a shared ancestor (apomorphies). Only derived characters, such as the spine-producing areoles of cacti, provide evidence for descent from a common ancestor. The results of cladistic analyses are expressed as cladograms: tree-like diagrams showing the pattern of evolutionary branching and descent. From the 1990s onwards, the predominant approach to constructing phylogenies for living plants has been molecular phylogenetics, which uses molecular characters, particularly DNA sequences, rather than morphological characters like the presence or absence of spines and areoles. The difference is that the genetic code itself is used to decide evolutionary relationships, instead of being used indirectly via the characters it gives rise to. Clive Stace describes this as having "direct access to the genetic basis of evolution." As a simple example, prior to the use of genetic evidence, fungi were thought either to be plants or to be more closely related to plants than animals. Genetic evidence suggests that the true evolutionary relationship of multicelled organisms is as shown in the cladogram below – fungi are more closely related to animals than to plants.

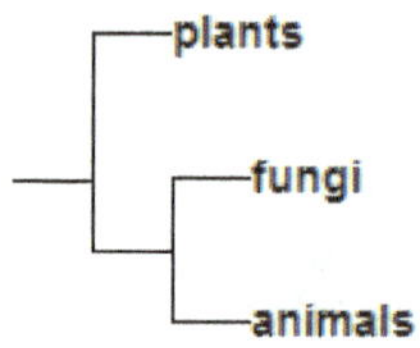

In 1998 the Angiosperm Phylogeny Group published a phylogeny for flowering plants based on an analysis of DNA sequences from most families of flowering plants. As a result of this work, many questions, such as which families represent the earliest branches of angiosperms, have now been answered. Investigating how plant species are related to each other allows botanists to better understand the process of evolution in plants. Despite the study of model plants and increasing use of DNA evidence, there is ongoing work and discussion among taxonomists about how best to classify plants into various taxa. Technological developments such as computers and electron microscopes have greatly increased the level of detail studied and speed at which data can be analysed.

REFERENCES

1. Liddell & Scott 1940.
2. Gordh & Headrick 2001, p. 134.
3. Online Etymology Dictionary 2012.
4. Bold 1977, p. 7.
5. Judd et al. 2002, p. 2.
6. Sumner 2000, p. 16.
7. Delcourt, Paul A.; Delcourt, Hazel R.; Cridlebaugh, Patricia A.; Chapman, Jefferson (1986-05-01). "Holocene ethnobotanical and paleoecological record of human impact on vegetation in the Little Tennessee River Valley, Tennessee". Quaternary Research. *25* (3): 330–349. Bibcode:1986QuRes..25..330D. doi:10.1016/0033-5894(86)90005-0.
8. Reed 1942, pp. 7–29.
9. Oberlies 1998, p. 155.

10. Needham, Lu & Huang 1986.
11. Greene 1909, pp. 140–142.
12. Bennett & Hammond 1902, p. 30.
13. Mauseth 2003, p. 532.
14. Dallal 2010, p. 197.
15. Panaino 2002, p. 93.
16. Levey 1973, p. 116.
17. Hill 1915.
18. National Museum of Wales 2007.
19. Yaniv & Bachrach 2005, p. 157.
20. Sprague 1939.
21. Waggoner 2001.
22. Scharf 2009, pp. 73–117.
23. Capon 2005, pp. 220–223.
24. Hoek, Mann & Jahns 2005, p. 9.
25. Starr 2009, p. 299–.
26. Morton 1981, p. 377.
27. Harris 2000, pp. 76–81.
28. Small 2012, p. 118–.
29. Karp 2009, p. 382.
30. National Science Foundation 1989.
31. Chaffey 2007, pp. 481–482.
32. Tansley 1935, pp. 299–302.
33. Willis 1997, pp. 267–271.
34. Morton 1981, p. 457.
35. de Candolle 2006, pp. 9–25, 450–465.
36. Jasechko et al. 2013, pp. 347–350.
37. Nobel 1983, p. 608.
38. Yates & Mather 1963, pp. 91–129.
39. Finney 1995, pp. 554–573.
40. Cocking 1993.
41. Cousens & Mortimer 1995.
42. Ehrhardt & Frommer 2012, pp. 1–21.
43. Haberlandt 1902, pp. 69–92.
44. Leonelli et al. 2012.
45. Sattler & Jeune 1992, pp. 249-262.

46. Sattler 1992, pp. 708-714.
47. Ereshefsky 1997, pp. 493–519.
48. Gray & Sargent 1889, pp. 292–293.
49. Medbury 1993, pp. 14–16.
50. Judd et al. 2002, pp. 347–350.
51. Burger 2013.
52. Kress et al. 2005, pp. 8369–8374.
53. Janzen et al. 2009, pp. 12794–12797.
54. Campbell et al. 2008, pp. 186–187.
55. Campbell et al. 2008, p. 1240.
56. Gust 1996.
57. Missouri Botanical Garden 2009.
58. Chapman et al. 2001, p. 56.
59. Braselton 2013.
60. Ben-Menahem 2009, p. 5368.
61. Campbell et al. 2008, p. 602.
62. Campbell et al. 2008, pp. 619–620.
63. Capon 2005, pp. 10–11.
64. Mauseth 2003, pp. 1–3.
65. Cleveland Museum of Natural History 2012.
66. Campbell et al. 2008, pp. 516–517.
67. Botanical Society of America 2013.
68. Ben-Menahem 2009, pp. 5367–5368.
69. Butz 2007, pp. 534–553.
70. Stover & Simmonds 1987, pp. 106–126.
71. Zohary & Hopf 2000, pp. 20–22.
72. Floros, Newsome & Fisher 2010.
73. Schoening 2005.
74. Acharya & Anshu 2008, p. 440.
75. [b] Kuhnlein, Harriet V.; Turner, Nancy J. (1991-01-01). Traditional Plant Foods of Canadian Indigenous Peoples: Nutrition, Botany, and Use. Taylor & Francis. ISBN 9782881244650.
76. Lüttge 2006, pp. 7–25.
77. Campbell et al. 2008, pp. 190–193.
78. Kim & Archibald 2009, pp. 1–39
79. Howe et al. 2008, pp. 2675–2685.
80. Takaichi 2011, pp. 1101–1118.

81. Lewis & McCourt 2004, pp. 1535–1556.
82. Padmanabhan & Dinesh-Kumar 2010, pp. 1368–1380.
83. Schnurr et al. 2002, pp. 1700–1709.
84. Ferro et al. 2002, pp. 11487–11492.
85. Kolattukudy 1996, pp. 83–108.
86. Fry 1989, pp. 1–11.
87. Thompson & Fry 2001, pp. 23–34.
88. Kenrick & Crane 1997, pp. 33–39.
89. Gowik & Westhoff 2010, pp. 56–63.
90. Benderoth et al. 2006, pp. 9118–9123.
91. Jeffreys 2005, pp. 38–40.
92. Mann 1987, pp. 186–187.
93. University of Maryland Medical Center 2011.
94. Frances, Densmore (1974). How Indians Use Wild Plants for Food, Medicine, and Crafts. Dover Publications. ISBN 9780486131108.
95. McCutcheon, A. R.; Ellis, S. M.; Hancock, R. E.; Towers, G. H. (1992-10-01). "Antibiotic screening of medicinal plants of the British Columbian native peoples". Journal of Ethnopharmacology. *37* (3): 213–223. doi:10.1016/0378-8741(92)90036-q. ISSN 0378-8741. PMID 1453710.
96. Klemm et al. 2005.
97. Scharlemann & Laurance 2008, pp. 52–53.
98. Research confirms Native American use of sweetgrass as bug repellent". Washington Post. Retrieved 2016-05-05.
99. Mauseth 2003, pp. 786–818.
100. TeachEthnobotany (2012-06-12), Cultivation of peyote by Native Americans: Past, present and future, retrieved 2016-05-05
101. Burrows 1990, pp. 1–73.
102. Addelson 2003.
103. Grime & Hodgson 1987, pp. 283–295.
104. Mauseth 2003, pp. 819–848.
105. Campbell et al. 2008, p. 794.
106. Herrera & Pellmyr 2002, pp. 211–235.
107. Proctor & Yeo 1973, p. 479.
108. Herrera & Pellmyr 2002, pp. 157–185.
109. Herrera & Pellmyr 2002, pp. 185–210.
110. Bennett & Willis 2001, pp. 5–32.
111. Beerling, Osborne & Chaloner 2001, pp. 287–394.

112. Björn et al. 1999, pp. 449–454.
113. Ben-Menahem 2009, pp. 5369–5370.
114. Ben-Menahem 2009, p. 5369.
115. Stace 2010b, pp. 629–633.
116. Hancock 2004, pp. 190–196.
117. Sobotka, Sáková & Curn 2000, pp. 103–112.
118. Renner & Ricklefs 1995, pp. 596–606.
119. Porley & Hodgetts 2005, pp. 2–3.
120. Savidan 2000, pp. 13–86.
121. Campbell et al. 2008, pp. 495–496.
122. Morgensen 1996, pp. 383–384.
123. Arabidopsis Genome Initiative 2000, pp. 796–815.
124. Devos & Gale 2000.
125. University of California-Davis 2012.
126. Russin et al. 1996, pp. 645–658.
127. Rochaix, Goldschmidt-Clermont & Merchant 1998, p. 550.
128. Glynn et al. 2007, pp. 451–461.
129. Possingham & Rose 1976, pp. 295–305.
130. Sun et al. 2002, pp. 95–100.
131. Heinhorst & Cannon 1993, pp. 1–9.
132. Schell & Van Montagu 1977, pp. 159–179.
133. Bird 2007, pp. 396–398.
134. Hunter 2008.
135. Spector 2012, p. 8.
136. Reik 2007, pp. 425–432.
137. Costa & Shaw 2007, pp. 101–106.
138. Mauseth 2003, pp. 552–581.
139. Copeland 1938, pp. 383–420.
140. Woese et al. 1977, pp. 305–311.
141. Cavalier-Smith 2004, pp. 1251–1262.
142. Mauseth 2003, pp. 617–654.
143. Becker & Marin 2009, pp. 999–1004.
144. Fairon-Demaret 1996, pp. 217–233.
145. Stewart & Rothwell 1993, pp. 279–294.
146. Taylor, Taylor & Krings 2009, chapter 13.
147. Mauseth 2003, pp. 720–750.

148. Mauseth 2003, pp. 751–785.
149. Lee et al. 2011, p. e1002411.
150. Mauseth 2003, pp. 278–279.
151. Mauseth 2003, pp. 280–314.
152. Mauseth 2003, pp. 315–340.
153. Mauseth 2003, pp. 341–372.
154. Mauseth 2003, pp. 373–398.
155. Mauseth 2012, p. 351.
156. Darwin 1880, pp. 129–200.
157. Darwin 1880, pp. 449–492.
158. Darwin 1880, p. 573.
159. Plant Hormones 2013.
160. Went & Thimann 1937, pp. 110–112.
161. Mauseth 2003, pp. 411–412.
162. Sussex 2008, pp. 1189–1198.
163. Campbell et al. 2008, pp. 827–830.
164. Mauseth 2003, pp. 411–413.
165. Taiz & Zeiger 2002, pp. 461–492.
166. Taiz & Zeiger 2002, pp. 519–538.
167. Lin, Zhong & Grierson 2009, pp. 331–336.
168. Taiz & Zeiger 2002, pp. 539–558.
169. Demole, Lederer & Mercier 1962, pp. 675–685.
170. Chini et al. 2007, pp. 666–671.
171. Roux 1984, pp. 25–29.
172. Raven, Evert & Eichhorn 2005, p. 9.
173. Mauseth 2003, pp. 433–467.
174. National Center for Biotechnology Information 2004.
175. Mauseth 2003, pp. 62–81.
176. López-Bautista, Waters & Chapman 2003, pp. 1715–1718.
177. Campbell et al. 2008, pp. 630, 738.
178. Campbell et al. 2008, p. 739.
179. Campbell et al. 2008, pp. 607–608.
180. Lepp 2012.
181. Campbell et al. 2008, pp. 812–814.
182. Campbell et al. 2008, p. 740.
183. Mauseth 2003, pp. 185–208.

184. Mithila et al. 2003, pp. 408–414.
185. Campbell et al. 2008, p. 741.
186. Mauseth 2003, pp. 114–153.
187. Mauseth 2003, pp. 154–184.
188. Capon 2005, p. 11.
189. Mauseth 2003, pp. 209–243.
190. Mauseth 2003, pp. 244–277.
191. Sattler & Jeune 1992, pp. 249-269.
192. Lilburn et al. 2006.
193. McNeill et al. 2011, p. Preamble, para. 7.
194. Mauseth 2003, pp. 528–551.
195. Mauseth 2003, pp. 528–55.
196. International Association for Plant Taxonomy 2006.
197. Silyn-Roberts 2000, p. 198.
198. Mauseth 2012, pp. 438–444.
199. Mauseth 2012, pp. 446–449.
200. Anderson 2001, pp. 26–27.
201. Mauseth 2012, pp. 442–450.
202. Stace 2010a, p. 104.
203. Mauseth 2012, p. 453.
204. Chase et al. 2003, pp. 399–436.
205. Capon 2005, p. 223.
206. Morton 1981, pp. 459–459.

CHAPTER 2

Plants Group

Chapter Contents

- **Introduction**
- **Archaeplastida**
- **Bryophyte**
- **Life Cycle**
- **Non-Vascular Plant**
- **Spermatophyte**
- **Pteridophyte**
- **Gymnosperms**
- **Flowering Plants**

Plants are more difficult to put into groups for many people possibly as they are so completely unlike us. Even a fly has a head, eyes and legs that we can identify with to some degree, but where do you start with a plant?

Add to this that the great majority of plants that we encounter belong to the same group and that the others are relatively uncommon with the differences being not that obvious until you go looking much more carefully and it's easy to see why most people's idea of plant groups are flowers/not-flowers or even edible/non-edible.

INTRODUCTION

Plants, also called green plants, are multicellular eukaryotes of the kingdom Plantae. They form an unranked cladeViridiplantae (Latin for green plants) that includes the flowering plants, conifers and other gymnosperms, ferns, clubmosses, hornworts, liverworts, mosses and the green algae. Green plants exclude the red and brown algae, the fungi, archaea, bacteria and animals.

Green plants have cell walls with cellulose and obtain most of their energy from sunlight via photosynthesis by primary chloroplasts, derived from endosymbiosis with cyanobacteria. Their chloroplasts contain chlorophylls a and b, which gives them their green color. Some plants are parasitic and have lost the ability to produce normal amounts of chlorophyll or to photosynthesize. Plants are characterized by sexual reproduction and alternation of generations, although asexual reproduction is also common.

There are about 300–315 thousand species of plants, of which the great majority, some 260–290 thousand, are seed plants (see the table below). Green plants provide most of the world's molecular oxygen and are the basis of most of Earth's ecologies, especially on land. Plants that produce grains, fruits and vegetables form humankind's basic foodstuffs, and have been domesticated for millennia. Plants play many roles in culture. They are used as ornaments and, until recently and in great variety, they have served as the source of most medicines and drugs. The scientific study of plants is known as botany, a branch of biology. Plants are one of the two groups into which all living things were traditionally divided; the other is animals. The division goes back at least as far as Aristotle (384 BC – 322 BC), who distinguished between plants, which generally do not move, and animals, which often are mobile to catch their food. Much later, when Linnaeus (1707–1778) created the basis of the modern system of scientific classification, these two groups became the kingdoms Vegetabilia (later Metaphyta or Plantae) and Animalia (also called Metazoa). Since then, it has become clear that the plant kingdom as originally defined included several unrelated groups, and the fungi and several groups of algae were removed to new kingdoms. However, these organisms are still often considered plants, particularly in popular contexts. Outside of formal scientific contexts, the term "plant" implies an association with certain traits, such as being multicellular, possessing cellulose, and having the ability to carry out photosynthesis.

Current Definitions of Plantae

When the name Plantae or plant is applied to a specific group of organisms or taxon, it usually refers to one of four concepts. From least to most inclusive, these four groupings are:

Name(s)	Scope	Description
Land plants, also known as Embryophyta	Plantae *sensu strictissimo*	This group includes the liverworts, hornworts, mosses, and vascular plants, as well as fossil plants similar to these surviving groups (e.g., Metaphyta Whittaker, 1969, Plantae Margulis, 1971).
Green plants, also known as **Viridiplantae**, **Viridiphyta** or **Chlorobionta**	Plantae *sensu stricto*	This group includes the green algae, and land plants that emerged within them, including stoneworts. The names given to these groups vary considerably as of July 2011. Viridiplantae encompass a group of organisms that have cellulose in their cell walls, possess chlorophylls *a* and *b* and have plastids that are bound by only two membranes that are capable of storing starch. It is this clade that is mainly the subject of this article (e.g., Plantae Copeland, 1956).

Archaeplastida, Plastida or Primoplantae	Plantae *sensu lato*	This group comprises the green plants above plus Rhodophyta (red algae) and Glaucophyta (glaucophyte algae). This clade includes the organisms that eons ago acquired their chloroplasts directly by engulfing cyanobacteria (e.g., Plantae Cavalier-Smith, 1981).
Old definitions of plant	Plantae *sensu amplo*	Old classifications placed diverse algae, fungi or bacteria in Plantae (e.g., Plantae or Vegetabilia Linnaeus, Plantae Haeckel 1866, Metaphyta Haeckel, 1894, Plantae Whittaker, 1969).

Another way of looking at the relationships between the different groups that have been called "plants" is through a cladogram, which shows their evolutionary relationships. The evolutionary history of plants is not yet completely settled, but one accepted relationship between the three groups described above is shown below. Those which have been called "plants" are in bold.

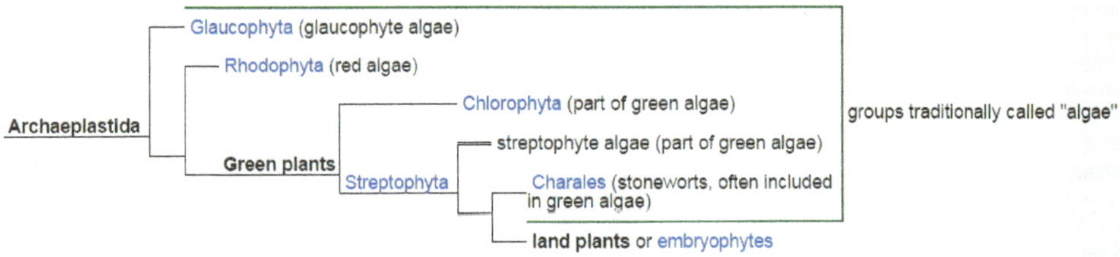

The way in which the groups of green algae are combined and named varies considerably between authors.

Algae

Algae comprise several different groups of organisms which produce energy through photosynthesis and for that reason have been included in the plant kingdom in the past. Most conspicuous among the algae are the seaweeds, multicellular algae that may roughly resemble land plants, but are classified among the brown, red and green algae. Each of these algal groups also includes various microscopic and single-celled organisms. There is good evidence that some of these algal groups arose independently from separate non-photosynthetic ancestors, with the result that many groups of algae are no longer classified within the plant kingdom as it is defined here. The Viridiplantae, the green plants – green algae and land plants – form a clade, a group consisting of all the descendants of a common ancestor. With a few exceptions among green algae, the green plants have the following features in common; cell walls containing cellulose, chloroplasts containing chlorophylls *a* and *b*, and food stores in the form of starch contained within plastids. They undergo closed mitosis without centrioles, and typically have mitochondria with flat cristae. The chloroplasts of green plants are surrounded by two membranes, suggesting they originated directly from endosymbiotic cyanobacteria.

Green algae from Ernst Haeckel's *Kunstformen der Natur*, 1904.

Two additional groups, the Rhodophyta (red algae) and Glaucophyta (glaucophyte algae), also have chloroplasts which appear to be derived directly from endosymbiotic cyanobacteria, although they differ in the pigments which are used in photosynthesis and so are different in colour. All three groups together are generally believed to have a single common origin, and so are classified together in the taxon Archaeplastida, whose name implies that the chloroplasts or plastids of all the members of the taxon were derived from a single ancient endosymbiotic event. This is the broadest modern definition of the plants.

In contrast, most other algae (e.g. brown algae/diatoms, haptophytes, dinoflagellates, and euglenids) not only have different pigments but also have chloroplasts with three or four surrounding membranes. They are not close relatives of the Archaeplastida, presumably having acquired chloroplasts separately from ingested or symbiotic green and red algae. They are thus not included in even the broadest modern definition of the plant kingdom, although they were in the past.

The green plants or Viridiplantae were traditionally divided into the green algae (including the stoneworts) and the land plants. However, it is now known that the land plants evolved from within a group of green algae, so that the green algae by themselves are a paraphyletic group, i.e. a group that excludes some of the descendants of a common ancestor. Paraphyletic groups are generally avoided in modern classifications, so that in recent treatments the Viridiplantae have been divided into two clades, the Chlorophyta and the Streptophyta (or Charophyta).

The Chlorophyta (a name that has also been used for *all* green algae) are the sister group to the group from which the land plants evolved. There are about 4,300 species of mainly marine organisms, both unicellular and multicellular. The latter include the sea lettuce, *Ulva*. The other group within the Viridiplantae are the mainly freshwater or terrestrial Streptophyta, which consists of the land plants together with the Charophyta, itself consisting of several groups of green algae such as the desmids and stoneworts. (The names have been used differently, e.g. Streptophyta to mean the group that excludes the land plants and Charophyta for the stoneworts alone or the stoneworts plus the land plants.) Streptophyte algae are either unicellular or form multicellular filaments, branched or unbranched. The genus *Spirogyra* is a filamentous streptophyte alga familiar to many, as it is often used in teaching and is one of the organisms responsible for the algal "scum" that pond-owners so dislike. The freshwater stoneworts strongly resemble land plants and are believed to be their closest relatives. Growing underwater, they consist of a central stalk with whorls of branchlets, giving them a superficial resemblance to horsetails, species of the genus *Equisetum*, which are true land plants.

ARCHAEPLASTIDA

The Archaeplastida (or Plantae *sensu lato*) are a major group of eukaryotes, comprising the red algae (Rhodophyta), the green algae, and the land plants, together with a small group of freshwater unicellular algae called glaucophytes. The chloroplasts of all these organisms are surrounded by two membranes, suggesting they developed directly from endosymbiotic cyanobacteria. In all other groups besides the amoeboid *Paulinella chromatophora*, the chloroplasts are surrounded by three or four membranes, suggesting they were acquired secondarily from red or green algae. The cells of the Archaeplastida typically lack centrioles and have mitochondria with flat cristae. They usually have a cell wall including cellulose, and food is stored in the form of starch. However, these characteristics are also shared with other eukaryotes. The main evidence that the Archaeplastida form a monophyletic group comes from genetic studies, which indicate their plastids probably had a single origin. This evidence is disputed. Based on results to date, alternative evolutionary scenarios to a single primary endosymbiosis is not possible to confirm or refute.Photosynthetic organisms with plastids of different origin (such as brown algae) do not belong to the Archaeplastida. The archaeplastidans fall into two main evolutionary lines. The red algae are pigmented with chlorophyll *a* and phycobiliproteins, like most cyanobacteria, and accumulate starch outside the chloroplasts. The green algae and land plants – together known as Viridiplantae (Latin for "green plants") or Chloroplastida – are pigmented with chlorophylls *a* and *b*, but lack phycobiliproteins, and starch is accumulated inside the chloroplasts. The glaucophytes have typical cyanobacterial pigments, and are unusual in retaining a cell wall within their plastids (called cyanelles).

Archaeplastida should not be confused with the older and obsolete name Archiplastideae, which refers to cyanobacteria and other groups of bacteria.

Taxonomy

The consensus in 2005, when the group consisting of the glaucophytes and red and green algae and land plants was named 'Archaeplastida', was that it was a clade, i.e. was monophyletic. Many studies published since then have provided evidence in agreement. Other studies, though, have suggested that the group is paraphyletic. To date, the situation appears unresolved, but a strong signal for Plantae (Archaeplastida) monophyly has been demonstrated in a recent study (with an enrichment of red algal genes). The assumption made here is that Archaeplastida is a valid clade. Various names have been given to the group. Some authors have simply referred to the group as plants or Plantae. However, the name Plantae is ambiguous, since it has also been applied to less inclusive clades, such as Viridiplantae and embryophytes. To distinguish, the larger group is sometimes known as Plantae *sensu lato* ("plants in the broad sense").

To avoid ambiguity, other names have been proposed. Primoplantae, which appeared in 2004, seems to be the first new name suggested for this group.

Another name applied to this node is Plastida, defined as the clade sharing "plastids of primary (direct prokaryote) origin [as] in *Magnolia virginiana* Linnaeus 1753".

Although many studies have suggested the Archaeplastida form a monophyletic group, a 2009 paper argues that they are in fact paraphyletic. The enrichment of novel red algal genes in a recent study demonstrates a strong signal for Plantae (Archaeplastida) monophyly and an equally strong signal of gene sharing history between the red/green algae and other lineages. This study provides insight on how rich mesophilic red algal gene data are crucial for testing controversial issues in eukaryote evolution and for understanding the complex patterns of gene inheritance in protists.

The name Archaeplastida was proposed in 2005 by a large international group of authors (Adl *et al.*), who aimed to produce a classification for the eukaryotes which took into account morphology, biochemistry, and phylogenetics, and which had "some stability in the near term." They rejected the use of formal taxonomic ranks in favour of a hierarchical arrangement where the clade names do not signify rank. Thus, the phylum name 'Glaucophyta' and the class name 'Rhodophyceae' appear at the same level in their classification. The divisions proposed for the Archaeplastida are shown below in both tabular and diagrammatic form.

The glaucophyte*Glaucocystis*

Archaeplastida

- Glaucophyta Skuja, 1954 (Glaucocystophyta Kies & Kremer, 1986) – glaucophytes

Glaucophytes are a small group of freshwater single-celled algae. Their chloroplasts, called *cyanelles*, have a peptidoglycan layer, making them more similar to cyanobacteria than those of the remaining Archaeplastida.

The rhodophyte *Laurencia*

- Rhodophyceae Thuret, 1855, emend. Rabenhorst, 1863, emend. Adl *et al.*, 2005 (Rhodophyta Wettstein 1901) – red algae
- Red algae form one of the largest groups of algae. Most are seaweeds, being multicellular and marine. Their red colour comes from phycobiliproteins, used as accessory pigments in light capture for photosynthesis.
- Chloroplastida Adl *et al.*, 2005 (Viridiplantae Cavalier-Smith 1981; Chlorobionta Jeffrey 1982, emend. Bremer 1985, emend. Lewis and McCourt 2004; Chlorobiota Kendrick and Crane 1997)

Chloroplastida is the term chosen by Adl *et al.* for the group made up of the green algae and land plants (embryophytes). Except where lost secondarily, all have chloroplasts without a peptidoglycan layer and lack phycobiliproteins.

The chlorophyte*Stigeoclonium*

- Chlorophyta Pascher, 1914, emend. Lewis & McCourt, 2004 – green algae (part)

Adl et al. employ a narrow definition of the Chlorophyta; other sources include the Chlorodendrales and Prasinophytae, which may themselves be combined.

- Ulvophyceae Mattox & Stewart, 1984
- Trebouxiophyceae Friedl, 1995 (Pleurastrophyceae Mattox et al. 1984; Microthamniales Melkonian 1990)
- Chlorophyceae Christensen, 1994
- Chlorodendrales Fritsch, 1917 – green algae (part)

- Prasinophytae Cavalier-Smith, 1998, emend. Lewis & McCourt, 2004 – green algae (part)
- *Mesostigma* Lauterborn, 1894, emend. McCourt *in* Adl *et al.*, 2005 (Mesostigmata Turmel, Otis, and Lemieux 2002)
- Charophyta Karol *et al.*, 2001, emend. Lewis & McCourt, 2004 (Charophyceae Smith 1938, emend. Mattox and Stewart 1984) – green algae (part) and land plants

Charophyta *sensu lato*, as used by Adl *et al.*, is a monophyletic group which is made up of some green algae, including the stoneworts (Charophyta *sensu stricto*), as well as the land plants (embryophytes).

- Sub-divisions other than Streptophytina (below) were not given by Adl et al.

Other sources would include the green algal groups Chlorokybales, Klebsormidiales, Zygnematales and Coleochaetales.

- Streptophytina Lewis & McCourt, 2004 – stoneworts and land plants
- Charales Lindley 1836 (Charophytae Engler, 1887) – stoneworts
- Plantae Haeckel 1866 (Cormophyta Endlicher, 1836; Embryophyta Endlicher, 1836, emend. Lewis & McCourt, 2004) – land plants (embryophytes)

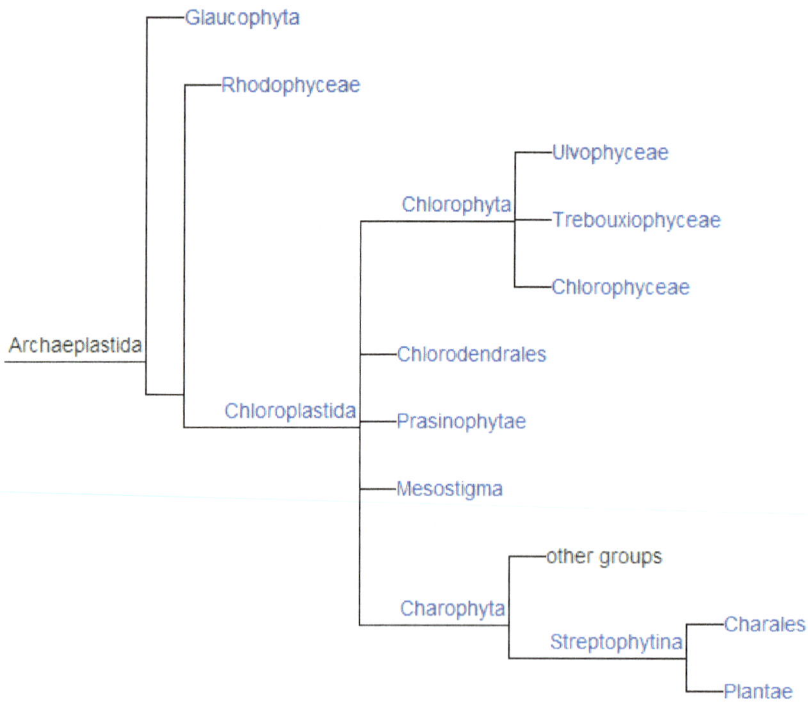

Morphology

All archaeplastidans have plastids (chloroplasts) that carry out photosynthesis and are believed to be derived from captured cyanobacteria. In glaucophytes, perhaps the most primitive members of the group, the chloroplast is called a *cyanelle* and shares several features with cyanobacteria, including a peptidoglycan cell wall, that are not retained in other members of the group. The resemblance of cyanelles to cyanobacteria supports the endosymbiotic theory.

The cells of most archaeplastidans have walls, commonly but not always made of cellulose. The Archaeplastida vary widely in the degree of their cell organization, from isolated cells to filaments to colonies to multi-celled organisms. The earliest were unicellular, and many groups remain so today. Multicellularity evolved separately in several groups, including red algae, ulvophyte green algae, and in the green algae that gave rise to stoneworts and land plants.

Endosymbiosis

Because the ancestral archaeplastidan is hypothesized to have acquired its chloroplasts directly by engulfing cyanobacteria, the event is known as a *primary endosymbiosis* (as reflected in the name chosen for the group 'Archaeplastida' i.e. 'ancient plastid'). Evidence for primary endosymbosis includes the presence of a double membrane around the chloroplasts; one membrane belonged to the bacterium, and the other to the eukaryote that captured it. Over time, many genes from the chloroplast have been transferred to the nucleus of the host cell. The presence of such genes in the nuclei of eukaryotes without chloroplasts suggests this transfer happened early in the evolution of the group.

Other eukaryotes with chloroplasts appear to have gained them by engulfing a single-celled archaeplastidan with its own bacterially-derived chloroplasts. Because these events involve endosymbiosis of cells that have their own endosymbionts, the process is called *secondary endosymbiosis*. The chloroplasts of such eukaryotes are typically surrounded by more than two membranes, reflecting a history of multiple engulfment. The chloroplasts of euglenids and chlorarachniophytes appear to be captured green algae, whereas those of other photosynthetic eukaryotes, such as heterokont algae, cryptophytes, haptophytes, and dinoflagellates, appear to be captured red algae.

Fossil Record

Perhaps the most ancient remains of Archaeplastida are microfossils from the Roper group in northern Australia. The structure of these single-celled fossils resembles that of modern green algae. They date to the Mesoproterozoic Era, about 1500 to 1300 Ma (million years ago). These fossils are consistent with a molecular clock study that calculated that this clade diverged about 1500 Ma. The oldest fossil that can be assigned to a specific modern group is the red alga *Bangiomorpha*, from 1200 Ma. In the late Neoproterozoic Era, algal fossils became more numerous and diverse. Eventually, in the Paleozoic Era, plants emerged onto land, and have continued to flourish up to the present.

BRYOPHYTE

Bryophyte is a traditional name used to refer to all embryophytes (land plants) that are non-vascular plants, namely the mosses, hornworts, and liverworts. The defining features of bryophytes are that their life cycles are dominated by the gametophyte stage, that their sporophytes are unbranched and that they do not have a true vascular tissue containing lignin, although some have specialized tissues for the transport of water. Bryophytes are usually considered to be a paraphyletic group and not a natural (monophyletic) group, although some studies have produced contrary results. Regardless of their status, the name is convenient and remains in use as an informal collective term. Bryophytes produce enclosed reproductive structures (gametangia and sporangia), but they produce neither flowers nor seeds, reproducing via spores. The term "bryophyte" comes from Greek βρύον, *bryon*, "tree-moss, oyster-green" + φυτόν – *phyton* "plant".

Marchantia, an example of a liverwort.

LIFE CYCLE

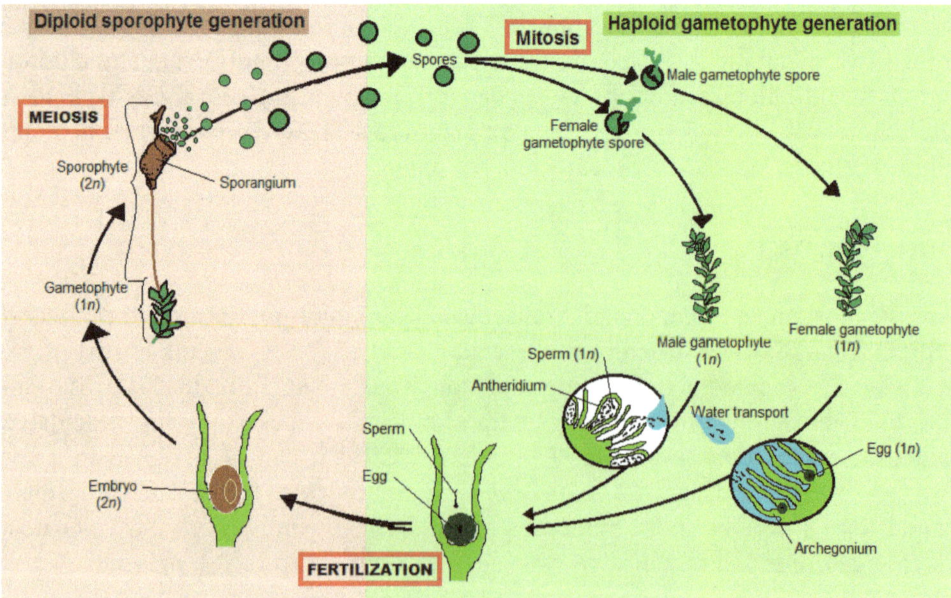

The life cycle of a dioicous bryophyte. The gametophyte (haploid) structures are shown in green, the sporophyte (diploid) in brown.

Like all land plants (embryophytes), bryophytes have life cycles with alternation of generations. In each cycle, a haploid gametophyte, each of whose cells contains a fixed number of unpaired chromosomes, alternates with a diploid sporophyte, whose cell contain two sets of paired chromosomes. Gametophytes produce haploid sperm and eggs which fuse to form diploid zygotes that grow into sporophytes. Sporophytes produce haploid spores by meiosis, that grow into gametophytes.

Bryophytes are gametophyte dominant, meaning that the more prominent, longer-lived plant is the haploid gametophyte. The diploid sporophytes appear only occasionally and remain attached to and nutritionally dependent on the gametophyte. In bryophytes, the sporophytes are always unbranched and produce a single sporangium (spore producing capsule). Liverworts, mosses and hornworts spend most of their lives as gametophytes. Gametangia (gamete-producing organs), archegonia and antheridia, are produced on the gametophytes, sometimes at the tips of shoots, in the axils of leaves or hidden under thalli. Some bryophytes, such as the liverwort *Marchantia*, create elaborate structures to bear the gametangia that are called gametangiophores. Sperm are flagellated and must swim from the antheridia that produce them to archegonia which may be on a different plant. Arthropods may assist in transfer of sperm. Fertilized eggs become zygotes, which develop into sporophyte embryos inside the archegonia. Mature sporophytes remain attached to the gametophyte. They consist of a stalk called a seta and a single sporangium or capsule. Inside the sporangium, haploid spores are produced by meiosis. These are dispersed, most commonly by wind, and if they land in a suitable environment can develop into a new gametophyte. Thus bryophytes disperse by a combination of swimming sperm and spores, in a manner similar to lycophytes, ferns and other cryptogams.

Sexuality

The arrangement of antheridia and archegonia on an individual bryophyte plant is usually constant within a species, although in some species it may depend on environmental conditions. The main division is between species in which the antheridia and archegonia occur on the same plant and those in which they occur on different plants. The term **monoicous** may be used where antheridia and archegonia occur on the same gametophyte and the term **dioicous** where they occur on different gametophytes. In seed plants, "monoecious" is used where flowers with anthers (microsporangia) and flowers with ovules (megasporangia) occur on the same sporophyte and "dioecious" where they occur on different sporophytes. These terms occasionally may be used instead of "monoicous" and "dioicous" to describe bryophyte gametophytes. "Monoecious" and "monoicous" are both derived from the Greek for "one house", "dioecious" and "dioicous" from the Greek for two houses. The use of the "oicy" terminology is said to have the advantage of emphasizing the difference between the gametophyte sexuality of bryophytes and the sporophyte sexuality of seed plants. Monoicous plants are necessarily **bisexual** (or hermaphroditic), meaning that the same plant has both sexes. The exact arrangement of the antheridia and archegonia in monoicous plants varies. They may be borne on different shoots (autoicous or autoecious), on the same shoot but not together in a common structure (paroicous or paroecious), or together in a common "inflorescence" (synoicous or synoecious). Dioicous plants are **unisexual**, meaning that the same plant has only one sex. All four patterns (autoicous, paroicous, synoicous and dioicous) occur in species of the moss genus *Bryum*.

Classification and Phylogeny

Hornworts include those bryophytes that are believed to be the closest living relatives of the vascular plants.

Mosses are one group of bryophytes.

Traditionally, all living land plants without vascular tissues were classified in a single taxonomic group, often a division (or phylum). More recently, phylogenetic research has questioned whether the bryophytes form a monophyletic group and thus whether they should form a single taxon. Although a 2005 study supported the traditional view that the bryophytes form a monophyletic group, by 2010 a broad consensus had emerged among systematists that bryophytes as a whole are not a natural group (i.e., are paraphyletic), although each of the three extant (living) groups is monophyletic.

The three bryophyte clades are the Divisions Marchantiophyta (liverworts), Bryophyta (mosses) and Anthocerotophyta (hornworts). The vascular plants or tracheophytes form a fourth, unranked clade of land plants called the "Polysporangiophyta". In this analysis, hornworts are sister to vascular plants and liverworts are sister to all other land plants, including the hornworts and mosses. Phylogenetic studies continue to produce conflicting results. In particular those based on gene sequences suggest the bryophytes are paraphyletic, whereas those based on the amino acid translations of the same genes suggest they are monophyletic. A 2014 study concluded that composition biases were responsible for these differences and that the bryophytes are monophyletic. The issue remains unresolved.

Paraphyletic View

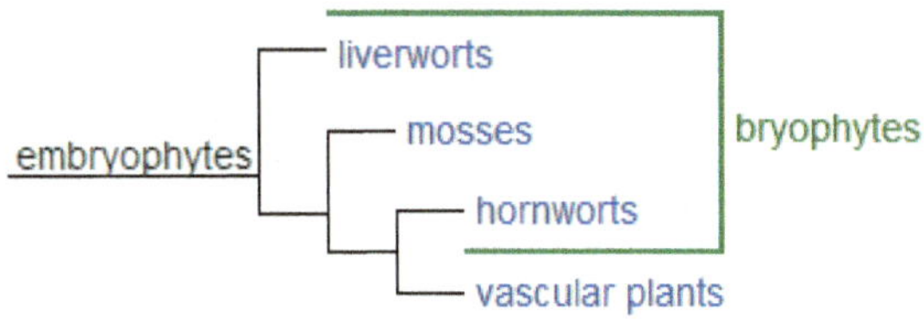

When extinct plants are taken into account, the picture is slightly altered. Some extinct land plants, such as the horneophytes, are not bryophytes, but also are not vascular plants because, like bryophytes, they do not have true vascular tissue. A different distinction is needed. In bryophytes, the sporophyte is a simple unbranched structure with a single spore-forming organ (sporangium). In all other land plants, the polysporangiophytes, the sporophyte is branched and carries many sporangia. It has been argued that this contrast between bryophytes and other land plants is less misleading than the traditional one of non-vascular versus vascular plant, since many mosses have well-developed water-conducting vessels. The contrast is shown in a slightly different cladogram:

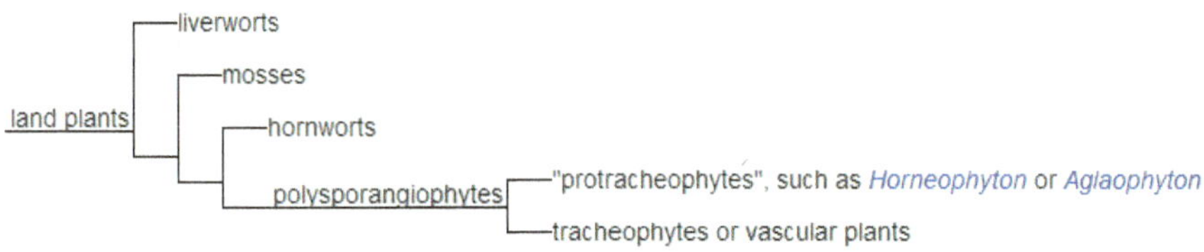

The term "bryophyte" thus refers to a grade of lineages defined primarily by what they lack. Compared to other living land plants, they lack vascular tissue containing lignin and branched sporophytes bearing multiple sporangia. The prominence of the gametophyte in the life cycle is also a shared feature of the three bryophyte lineages (extant vascular plants are all sporophyte dominant).

Other Views

An alternative phylogeny, based on amino acids rather than genes, shows bryophytes as a monophyletic group:

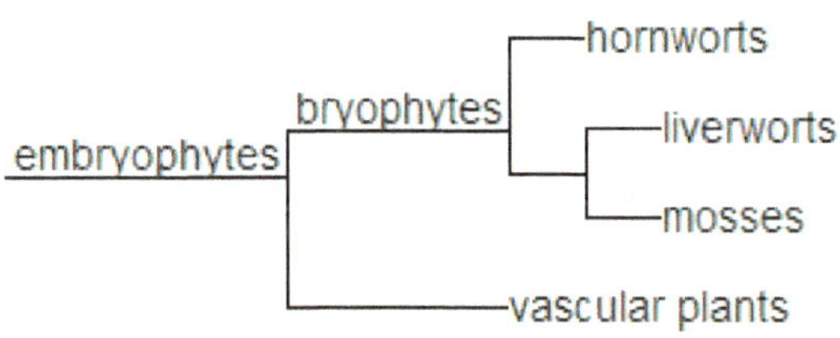

If this phylogeny proves correct, then the complex sporophyte of living vascular plants might have evolved independently of the simpler unbranched sporophyte present in bryophytes. Other studies suggest a monophyletic group comprising liverworts and mosses, with hornworts being sister to vascular plants.

Evolution

Bryophytes are the oldest of all lineages of land plants and are believed to be a vital link in the migration of plants from aquatic environments onto land. A number of physical features link bryophytes to both land plants and aquatic plants. Two distinct adaptations have helped to make the move from water to land possible and forged the way for plants to colonize the Earth's terrestrial environments. A waxy cuticle covering the soft tissue of the plant provides protection and prevents desiccation of the plant's tissues. The development of gametangia provided further protection specifically for gametes.

They also have embryonic development which is a significant adaptation seen in land plants and not green algae. Connections to their aquatic ancestry are also evident through their dependence on water for reproduction and survival. A thin layer of water is required on the surface of the plant to enable the movement of sperm between gametophytes and the fertilization of an egg.

Comparative Morphology

Summary of the morphological characteristics of the gametophytes of the three groups of bryophytes:

	Liverworts	Mosses	Hornworts
Structure	Thalloid or Foliose	Foliose	Thalloid
Symmetry	Dorsiventral or radial	Radial	Dorsiventral
Rhizoids	Unicellular	Pluricellular	Unicellular
Chloroplasts/cell	Many	Many	One
Protonemata	Reduced	Present	Absent
Gametangia (antheridia and archegonia)	Superficial	Superficial	Immersed

Summary of the morphological characteristics of the sporophytes of the three groups of bryophytes:

	Liverworts	Mosses	Hornworts
Stomata	Absent	Present	Present

Structure	Small, without chlorophyll	Large, with chlorophyll	Large, with chlorophyll
Persistence	Ephemeral	Persistent	Persistent
Growth	Defined	Defined	Continuous
Seta	Present	Present	Absent
Capsule form	Simple	Differentiated (operculum, peristome)	Elongated
Maturation of spores	Simultaneous	Simultaneous	Graduate
Dispersion of spores	Elaters	Peristome teeth	Pseudo-elaters
Columella	Absent	Present	Present
Dehiscence	Longitudinal or irregular	Transverse	Longitudinal

NON-VASCULAR PLANT

Algae, a type of non-vascular plant

Non-vascular plants are plants without a vascular system consisting of xylem and phloem. Although non-vascular plants lack these particular tissues, many possess simpler tissues that are specialized for internal transport of water. Non-vascular plants do not have a wide variety of specialized tissue types. Mosses and leafy liverworts have structures that look like leaves, but are not true leaves because they are single sheets of cells with no stomata, no internal air spaces and have no xylem or phloem. Consequently, they are unable to control the rate of water loss from their tissues and are said to be poikilohydric. These organisms do, however, contain an elementary cuticle which was important in the evolution of land plants. All land plants have a life cycle with an alternation of generations between a diploid sporophyte and a haploid gametophyte, but in all non-vascular land plants the gametophyte generation is dominant. In these plants, the sporophytes grow from and are dependent on gametophytes for taking in water and mineral nutrients and for provision of photosynthate, the products of photosynthesis.

Non-vascular plants include two distantly related groups:

- Bryophytes, an informal group that is now treated as three separate land plant Divisions, namely Bryophyta (mosses), Marchantiophyta (liverworts), and Anthocerotophyta (hornworts). In all bryophytes, the primary plants are the haploid gametophytes, with the only diploid portion being the attached sporophyte, consisting of a stalk and sporangium. Because these plants lack lignified water-conducting tissues, they can't become as tall as most vascular plants.
- Algae - especially the green algae. Recent studies have demonstrated that the algae consist of several unrelated groups. It turns out that the common features of living in water and photosynthesis were misleading as indicators of close relationship. Only those groups of algae included in the Viridiplantae are still considered relatives of land plants.[6]

These groups are sometimes referred to as "lower plants", referring to their status as the earliest plant groups to evolve, but the usage is imprecise, since both groups are polyphyletic and may be used to include vascular cryptogams, such as the ferns and fern allies that reproduce using spores. Non-vascular plants are often among the first species to move into new and inhospitable territories, along with prokaryotes and protists, and thus function as pioneer species.

Vascular Plant

Vascular plants (from Latin *vasculum*: duct), also known as tracheophytes (from the equivalent Greek term *trachea*) and also higher plants, form a large group of plants (ca 308,312 accepted known species) that are defined as those land plants that have lignified tissues (the xylem) for conducting water and minerals throughout the plant. They also have a specialized non-lignified tissue (the phloem) to conduct products of photosynthesis. Vascular plants include the clubmosses, horsetails, ferns, gymnosperms (including conifers) and angiosperms (flowering plants). Scientific names for the group include Tracheophyta and Tracheobionta.

Characteristics

Vascular plants are distinguished by two primary characteristics:

- Vascular plants have vascular tissues which distribute resources through the plant. This feature allows vascular plants to evolve to a larger size than non-vascular plants, which lack these specialized conducting tissues and are therefore restricted to relatively small sizes.
- In vascular plants, the principal generation phase is the *sporophyte*, which is usually diploid with two sets of chromosomes per cell. Only the germ cells and gametophytes are haploid. By contrast, the principal generation phase in non-vascular plants is the *gametophyte*, which is haploid with one set of chromosomes per cell. In these plants, only the spore stalk and capsule are diploid.

One possible mechanism for the presumed switch from emphasis on the haploid generation to emphasis on the diploid generation is the greater efficiency in spore dispersal with more complex diploid structures. In other words, elaboration of the spore stalk enabled the production of more spores, and enabled the development of the ability to release them higher and to broadcast them farther. Such developments may include more photosynthetic area for the spore-bearing structure, the ability to grow independent roots, woody structure for support, and more branching.

Phylogeny

A proposed phylogeny of the vascular plants after Kenrick and Crane is as follows, with modification to the gymnosperms from Christenhusz *et al.* (2011a), Pteridophyta from Smith *et al.* and lycophytes and ferns by Christenhusz *et al.* (2011b)

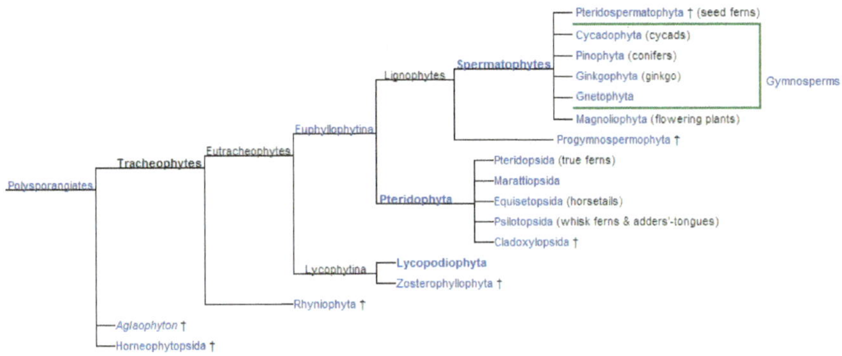

This phylogeny is supported by several molecular studies. Other researchers state that taking fossils into account leads to different conclusions, for example that the ferns (Pteridophyta) are not monophyletic.

Nutrient Distribution

Water and nutrients in the form of inorganic solutes are drawn up from the soil by the roots and transported throughout the plant by the xylem. Organic compounds such as sucrose produced by photosynthesis in leaves are distributed by the phloem sieve tube elements.

Photographs showing xylem elements in the shoot of a fig tree (*Ficus alba*): crushed in hydrochloric acid, between slides and cover slips.

The xylem consists of vessels in flowering plants and tracheids in other vascular plants, which are dead hard-walled hollow cells arranged to form files of tubes that function in water transport. A tracheid cell wall usually contains the polymer lignin. The phloem however consists of living cells called sieve-tube members. Between the sieve-tube members are sieve plates, which have pores to allow molecules to pass through. Sieve-tube members lack such organs as nuclei or ribosomes, but cells next to them, the companion cells, function to keep the sieve-tube members alive.

Transpiration

The most abundant compound in all plants, as in all cellular organisms, is water which serves an important structural role and a vital role in plant metabolism. Transpiration is the main process of water movement within plant tissues. Water is constantly transpired from the plant through its stomata to the atmosphere and replaced by soil water taken up by the roots. The movement of water out of the leaf stomata creates a transpiration pull or tension in the water column in the xylem vessels or tracheids. The pull is the result of water surface tension within the cell walls of the mesophyll cells, from the surfaces of which evaporation takes place when the stomata are open. Hydrogen bonds exist between water molecules, causing them to line up; as the molecules at the top of the plant evaporate, each pulls the next one up to replace it, which in turn pulls on the next one in line. The draw of water upwards may be entirely passive and can be assisted by the movement of water into the roots via osmosis. Consequently, transpiration requires very little energy to be used by the plant. Transpiration assists the plant in absorbing nutrients from the soil as soluble salts.

Absorption

Living root cells passively absorb water in the absence of transpiration pull via osmosis creating root pressure. It is possible for there to be no evapotranspiration and therefore no pull of water towards the shoots and leaves. This is usually due to high temperatures, high humidity, darkness or drought.

Conduction

Xylem and phloem tissues are involved in the conduction processes within plants. Sugars are conducted throughout the plant in the phloem, water and other nutrients through the xylem. Conduction occurs from a source to a sink for each separate nutrient. Sugars are produced in the leaves (a source) by photosynthesis and transported to the growing shoots and roots (sinks) for use in growth, cellular respiration or storage. Minerals are absorbed in the roots (a source) and transported to the shoots to allow cell division and growth.

SPERMATOPHYTE

The spermatophytes (from the Greek word Σπερματόφυτα), also known as phanerogams or phenogamae, comprise those plants that produce seeds, hence the alternative name seed plants. They are a subset of the embryophytes or land plants. The term *phanerogams* or *phanerogamae* is derived from the Greek φανερός, *phanerós* meaning «visible», in contrast to the cryptogamae from Greek κρυπτός *kryptós* = «hidden» together with the suffix γαμέω, *gameein*, "to marry".

These terms distinguished those plants with hidden sexual organs (cryptogamae) from those with visible sexual organs (phanerogamae).

Description

The living spermatophytes form five groups:

- cycads, a subtropical and tropical group of plants with a large crown of compound leaves and a stout trunk,
- *ginkgo*, a single living species of tree,
- conifers, cone-bearing trees and shrubs,
- gnetophytes, woody plants in the genera *Ephedra*, *Gnetum*, and *Welwitschia*
- angiosperms, (or magnoliophyta) the flowering plants, a large group including many familiar plants in a wide variety of habitats.

In addition to the taxa listed above, the fossil (old creature) record contains evidence of many extinct taxa of seed plants. The so-called "seed ferns" (Pteridospermae) were one of the earliest successful groups of land plants, and forests dominated by seed ferns were prevalent in the late Paleozoic. *Glossopteris* was the most prominent tree genus in the ancient southern supercontinent of Gondwana during the Permian period. By the Triassic period, seed ferns had declined in ecological importance, and representatives of modern gymnosperm groups were abundant and dominant through the end of the Cretaceous, when angiosperms radiated.

Evolution

A whole genome duplication event in the ancestor of seed plants occurred about 319 million years ago. This gave rise to a series of evolutionary changes that resulted in the origin of seed plants. A middle Devonian precursor to seed plants from Belgium has been identified predating the earliest seed plants by about 20 million years. *Runcaria*, small and radially symmetrical, is an integumented megasporangium surrounded by a cupule. The megasporangium bears an unopened distal extension protruding above the mutlilobed integument. It is suspected that the extension was involved in anemophilous pollination. *Runcaria* sheds new light on the sequence of character acquisition leading to the seed. *Runcaria* has all of the qualities of seed plants except for a solid seed coat and a system to guide the pollen to the seed.

Relationships and Nomenclature

Seed-bearing plants were traditionally divided into angiosperms, or flowering plants, and gymnosperms, which includes the gnetophytes, cycads, ginkgo, and conifers. Older morphological studies believed in a close relationship between the gnetophytes and the angiosperms, in particular based on vessel elements. However, molecular studies (and some more recent morphological and fossil papers) have generally shown a clade of gymnosperms, with the gnetophytes in or near the conifers. For example, one common proposed set of relationships is known as the *gne-pine hypothesis* and looks like:

However, the relationships between these groups should not be considered settled.

Other classifications group all the seed plants in a single division, with classes for our five groups:

- Division Spermatophyta
  - Cycadopsida, the cycads
  - Ginkgoopsida, the ginkgo
  - Pinopsida, the conifers, ("Coniferopsida")
  - Gnetopsida, the gnetophytes
  - Magnoliopsida, the flowering plants, or Angiospermopsida

A more modern classification ranks these groups as separate divisions (sometimes under the Superdivision Spermatophyta):

- Cycadophyta, the cycads
- Ginkgophyta, the ginkgo
- Pinophyta, the conifers
- Gnetophyta, the gnetophytes
- Magnoliophyta, the flowering plants

An alternative phylogeny of spermatophytes based on the work by Novíkov & Barabaš-Krasni 2015 with plant taxon authors from Anderson, Anderson & Cleal 2007 showing the relationship of extinct clades.

Unassigned Spermatophyta

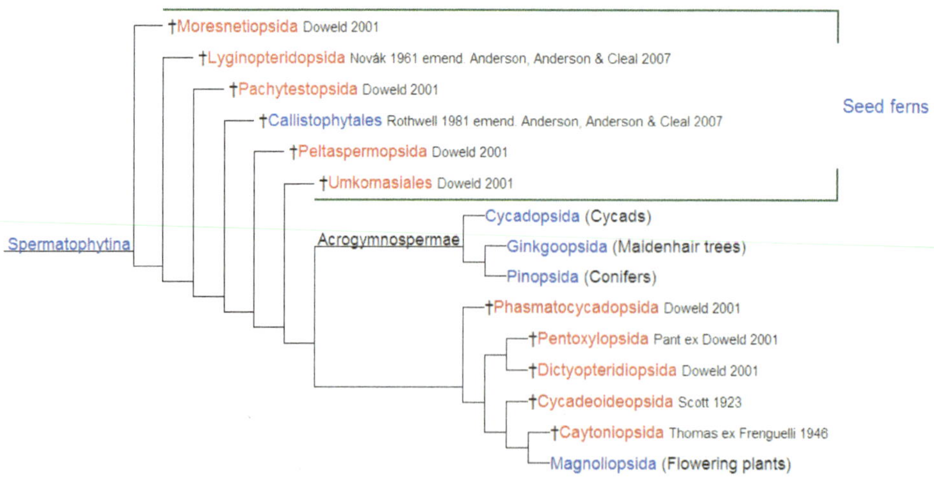

- Avatiaceae Anderson & Anderson 2003
- Axelrodiopsida Anderson & Anderson
- Alexiales Anderson & Anderson 2003

- Hamshawviales Anderson & Anderson 2003
- Hexapterospermales Doweld 2001
- Hlatimbiales Anderson & Anderson 2003
- Matatiellales Anderson & Anderson 2003
- Petriellales Taylor et al. 1994
- Arberiopsida Doweld 2001

PTERIDOPHYTE

Pteridophytes or **Pteridophyta**, in the broad interpretation of the term (or *sensu lato*), are vascular plants (plants with xylem and phloem) that reproduce and disperse via spores. Because they produce neither flowers nor seeds, they are referred to as cryptogams. The group includes ferns, horsetails, clubmosses, spikemosses and quillworts. These do not form a monophyletic group, because ferns and horsetails are more closely related to seed plants than to lycophytes (clubmosses, spikemosses and quillworts). Therefore, pteridophytes are no longer considered to form a valid taxon, but the term is still used as an informal way to refer to ferns (monilophytes) and lycophytes, and some recent authors have used the term to refer strictly to the monilophytes.

Athyrium filix-femina frond

Pteridophyte Classification

Pteridophytes consist of two separate classes:

- Lycopodiopsida
    - Lycopodiidae (clubmosses)
    - Selaginellidae (spikemosses, quillworts)
- Polypodiopsida
    - Psilotidae: Psilotales (whisk ferns) and Ophioglossales (e.g. grape ferns)

- Equisetidae (horsetails)
- Marattiidae (marattioid ferns)
- Polypodiidae (leptosporangiate ferns, the most species-rich group)

In addition to these living groups, several groups that are now extinct and known only from fossils are considered to belong to pteridophytes. These groups include the Rhyniophyta, Zosterophyllophyta, Trimerophytophyta, and the progymnosperms.

Modern studies of the land plants agree that all pteridophytes share a common ancestor, which is also the ancestor of seed plants. Therefore, pteridophytes do not form a clade but a paraphyletic group.

Pteridophyte life Cycle

Just as with seed plants and mosses, the life cycle of pteridophytes involves alternation of generations. This means that a diploid generation (the sporophyte, which produces spores) is followed by a haploid generation (the gametophyte or prothallus, which produces gametes). Pteridophytes differ from mosses and seed plants in that both generations are independent and free-living, although the sporophyte is generally much larger and more conspicuous. The sexuality of pteridophyte gametophytes can be classified as follows:

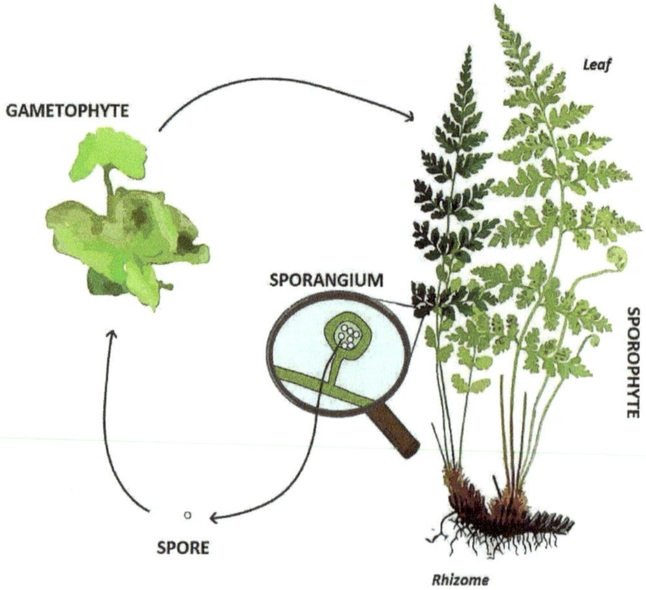

Pteridophyte life cycle

- Dioicous: each individual gametophyte is either male (producing antheridia and hence sperm) or female (producing archegonia and hence egg cells).

- Monoicous: each individual gametophyte produces both antheridia and archegonia and can function both as a male and as a female.

Protandrous: the antheridia mature before the archegonia (male first, then female).

Protogynous: the archegonia mature before the antheridia (female first, then male).

These terms are *not* the same as monoecious and dioecious, which refer to whether a seed plant sporophyte bears both male and female gametophytes (i.e. produces both pollen and seeds), or just one of the sexes.

GYMNOSPERMS

The **gymnosperms** are a group of seed-producing plants that includes conifers, cycads, *Ginkgo*, and gnetophytes. The term "gymnosperm" comes from the Greek composite word *γυμνόσπερμος* (γυμνός gymnos, naked and σπέρμα sperma, seed), meaning naked seeds, after the unenclosed condition of their seeds (called ovules in their unfertilized state). Their naked condition stands in contrast to the seeds and ovules of flowering plants (angiosperms), which are enclosed within an ovary. Gymnosperm seeds develop either on the surface of scales or leaves, often modified to form cones, or at the end of short stalks as in *Ginkgo*.

The gymnosperms and angiosperms together compose the spermatophytes or seed plants. By far the largest group of living gymnosperms are the conifers (pines, cypresses, and relatives), followed by cycads, gnetophytes (*Gnetum*, *Ephedra* and *Welwitschia*), and *Ginkgo* (a single living species).

Classification

In early classification schemes, the gymnosperms (Gymnospermae) were regarded as a "natural" group. There is conflicting evidence on the question of whether the living gymnosperms form a clade. The fossil record of gymnosperms includes many distinctive taxa that do not belong to the four modern groups, including seed-bearing trees that have a somewhat fern-like vegetative morphology (the so-called "seed ferns" or pteridosperms.) When fossil gymnosperms such as Bennettitales, *Caytonia* and the glossopterids are considered, it is clear that angiosperms are nested within a larger gymnosperm clade, although which group of gymnosperms is their closest relative remains unclear.

For the most recent classification on extant gymnosperms see Christenhusz *et al.* (2011). There are 12 families, 83 known genera with a total of ca 1080 known species (Christenhusz & Byng 2016).

Subclass Cycadidae

- Order Cycadales
    - Family Cycadaceae: Cycas
    - Family Zamiaceae: Dioon, Bowenia, Macrozamia, Lepidozamia, Encephalartos, Stangeria, Ceratozamia, Microcycas, Zamia.

Subclass Ginkgoidae

- Order Ginkgoales

  - Family Ginkgoaceae: Ginkgo

Subclass Gnetidae

- Order Welwitschiales
  - Family Welwitschiaceae: Welwitschia
- Order Gnetales
  - Family Gnetaceae: Gnetum
- Order Ephedrales
  - Family Ephedraceae: Ephedra

Subclass Pinidae

- Order Pinales
  - Family Pinaceae: Cedrus, Pinus, Cathaya, Picea, Pseudotsuga, Larix, Pseudolarix, Tsuga, Nothotsuga, Keteleeria, Abies
- Order Araucariales
  - Family Araucariaceae: Araucaria, Wollemia, Agathis
  - Family Podocarpaceae: Phyllocladus, Lepidothamnus, Prumnopitys, Sundacarpus, Halocarpus, Parasitaxus, Lagarostrobos, Manoao, Saxegothaea, Microcachrys, Pherosphaera, Acmopyle, Dacrycarpus, Dacrydium, Falcatifolium, Retrophyllum, Nageia, Afrocarpus, Podocarpus
- Order Cupressales
  - Family Sciadopityaceae: Sciadopitys
  - Family Cupressaceae: Cunninghamia, Taiwania, Athrotaxis, Metasequoia, Sequoia, Sequoiadendron, Cryptomeria, Glyptostrobus, Taxodium, Papuacedrus, Austrocedrus, Libocedrus, Pilgerodendron, Widdringtonia, Diselma, Fitzroya, Callitris (incl. Actinostrobus and Neocallitropsis), Thujopsis, Thuja, Fokienia, Chamaecyparis, Callitropsis, Cupressus, Juniperus, Xanthocyparis, Calocedrus, Tetraclinis, Platycladus, Microbiota
  - Family **Taxaceae**: *Austrotaxus*, *Pseudotaxus*, *Taxus*, *Cephalotaxus*, *Amentotaxus*, *Torreya*

Diversity and Origin

There are more than 1000 extant or currently living species of gymnosperms in 88 plant genera belonging to 14 plant families.

It is widely accepted that the gymnosperms originated in the late Carboniferous period, replacing the lycopsid rainforests of the tropical region. This appears to have been the result of a whole genome duplication event around 319 million years ago. Early characteristics of seed plants were evident in fossil progymnosperms of the late Devonian period around 380 million years ago. It has been suggested that during the mid-Mesozoic era, pollination of some extinct groups of gymnosperms was by extinct species of scorpionflies that had specialized proboscis for feeding on pollination drops. The scorpionflies likely engaged in pollination mutualisms with gymnosperms, long before the similar and independent coevolution of nectar-feeding insects on angiosperms. Evidence has also been found that mid-Mesozoic gymnosperms were pollinated by Kalligrammatid lacewings, a now-extinct genus with members which (in an example of convergent evolution) resembled the modern butterflies that arose far later.

Conifers are by far the most abundant extant group of gymnosperms with six to eight families, with a total of 65-70 genera and 600-630 species (696 accepted names). Conifers are woody plants and most are evergreens. The leaves of many conifers are long, thin and needle-like, other species, including most Cupressaceae and some Podocarpaceae, have flat, triangular scale-like leaves. *Agathis* in Araucariaceae and *Nageia* in Podocarpaceae have broad, flat strap-shaped leaves.

Cycads are the next most abundant group of gymnosperms, with two or three families, 11 genera, and approximately 300 species. The other extant groups are the 75-80 species of Gnetales and one species of Ginkgo.

Uses

Gymnosperms have major economic uses. Pine, fir, spruce, and cedar are all examples of conifers that are used for lumber. Some other common uses for gymnosperms are soap, varnish, nail polish, food, gum, and perfumes.

Life Cycle

Gymnosperms, like all vascular plants, have a sporophyte-dominant life cycle. The gametophyte (gamete-bearing phase) is relatively short-lived. Two spore types, microspores and megaspores, are typically produced in pollen cones or ovulate cones, respectively. Gametophytes, as with all heterosporous plants, develop within the spore wall. Pollen grains (microgametophytes) mature from microspores, and ultimately produce sperm cells. Megagametophytes develop from megaspores and are retained within the ovule. They typically produce multiple archegonia. During pollination, pollen grains are physically transferred between plants, from pollen cone to the ovule, being transferred by wind or insects. Whole grains enter each ovule through a microscopic gap in the ovule coat (integument) called the micropyle. The pollen grains mature further inside the ovule and produce sperm cells. Two main modes of fertilization are found in gymnosperms. Cycads and *Ginkgo* have motile sperm that swim directly to the egg inside the ovule, whereas conifers and gnetophytes have sperm with no flagella that are conveyed to the egg along a pollen tube. After syngamy (joining of the sperm and egg cell), the zygote develops into an embryo (young sporophyte). More than one embryo is usually initiated in each gymnosperm seed. The mature seed comprises the embryo and the remains of the female gametophyte, which serves as a food supply, and the seed coat (integument).

Genetics

The first published sequenced genome for any gymnosperm was the genome of *Picea abies* in 2013.

FLOWERING PLANTS

The **flowering plants (angiosperms)**, also known as **Angiospermae** or **Magnoliophyta**, are the most diverse group of land plants, with 416 families, approx. 13,164 known genera and a total of c. 295,383 known species. Like gymnosperms, angiosperms are seed-producing plants; they are distinguished from gymnosperms by characteristics including flowers, endosperm within the seeds, and the production of fruits that contain the seeds. Etymologically, angiosperm means a plant that produces seeds within an enclosure, in other words, a fruiting plant. The term "angiosperm" comes from the Greek composite word (*angeion*, "case" or "casing", and *sperma*, "seed") meaning "enclosed seeds", after the enclosed condition of the seeds.

The ancestors of flowering plants diverged from gymnosperms in the Triassic Period, during the range 245 to 202 million years ago (mya), and the first flowering plants are known from 160 mya. They diversified extensively during the Lower Cretaceous, became widespread by 120 mya, and replaced conifers as the dominant trees from 100 to 60 mya.

Angiosperm Derived Characteristics

Bud of a pink rose

Angiosperms differ from other seed plants in several ways, described in the table. These distinguishing characteristics taken together have made the angiosperms the most diverse and numerous land plants and the most commercially important group to humans.

Distinctive features of Angiosperms	
Feature	**Description**
Flowering organs	Flowers, the reproductive organs of flowering plants, are the most remarkable feature distinguishing them from the other seed plants. Flowers provided angiosperms with the means to have a more species-specific breeding system, and hence a way to evolve more readily into different species without the risk of crossing back with related species. Faster speciation enabled the Angiosperms to adapt to a wider range of ecological niches. This has allowed flowering plants to largely dominate terrestrial ecosystems.
Stamens with two pairs of pollen sacs	Stamens are much lighter than the corresponding organs of gymnosperms and have contributed to the diversification of angiosperms through time with adaptations to specialized pollination syndromes, such as particular pollinators. Stamens have also become modified through time to prevent self-fertilization, which has permitted further diversification, allowing angiosperms eventually to fill more niches.
Reduced male parts, three cells	The male gametophyte in angiosperms is significantly reduced in size compared to those of gymnosperm seed plants. The smaller size of the pollen reduces the amount of time between pollination — the pollen grain reaching the female plant — and fertilization. In gymnosperms, fertilization can occur up to a year after pollination, whereas in angiosperms, fertilization begins very soon after pollination. The shorter amount of time between pollination and fertilization allows angiosperms to produce seeds earlier after pollination than gymnosperms, providing angiosperms a distinct evolutionary advantage.
Closed carpel enclosing the ovules (carpel or carpels and accessory parts may become the fruit)	The closed carpel of angiosperms also allows adaptations to specialized pollination syndromes and controls. This helps to prevent self-fertilization, thereby maintaining increased diversity. Once the ovary is fertilized, the carpel and some surrounding tissues develop into a fruit. This fruit often serves as an attractant to seed-dispersing animals. The resulting cooperative relationship presents another advantage to angiosperms in the process of dispersal.

Reduced female gametophyte, seven cells with eight nuclei	The reduced female gametophyte, like the reduced male gametophyte, may be an adaptation allowing for more rapid seed set, eventually leading to such flowering plant adaptations as annual herbaceous life-cycles, allowing the flowering plants to fill even more niches.
Endosperm	In general, endosperm formation begins after fertilization and before the first division of the zygote. Endosperm is a highly nutritive tissue that can provide food for the developing embryo, the cotyledons, and sometimes the seedling when it first appears.

Evolution

Fossilized spores suggest that higher plants (embryophytes) have lived on land for at least 475 million years. Early land plants reproduced sexually with flagellated, swimming sperm, like the green algae from which they evolved. An adaptation to terrestrialization was the development of upright meiosporangia for dispersal by spores to new habitats. This feature is lacking in the descendants of their nearest algal relatives, the Charophycean green algae. A later terrestrial adaptation took place with retention of the delicate, avascular sexual stage, the gametophyte, within the tissues of the vascular sporophyte. This occurred by spore germination within sporangia rather than spore release, as in non-seed plants. A current example of how this might have happened can be seen in the precocious spore germination in *Selaginella*, the spike-moss. The result for the ancestors of angiosperms was enclosing them in a case, the seed. The first seed bearing plants, like the ginkgo, and conifers (such as pines and firs), did not produce flowers. The pollen grains (males) of *Ginkgo* and cycads produce a pair of flagellated, mobile sperm cells that "swim" down the developing pollen tube to the female and her eggs.

Flowers of *Malus sylvestris* (crab apple)

Flowers and leaves of *Oxalis pes-caprae* (Bermuda buttercup)

The apparently sudden appearance of nearly modern flowers in the fossil record initially posed such a problem for the theory of evolution that Charles Darwin called it an "*abominable mystery*". However, the fossil record has considerably grown since the time of Darwin, and recently discovered angiosperm fossils such as *Archaefructus*, along with further discoveries of fossil gymnosperms, suggest how angiosperm characteristics may have been acquired in a series of steps. Several groups of extinct gymnosperms, in particular seed ferns, have been proposed as the ancestors of flowering plants, but there is no continuous fossil evidence showing exactly how flowers evolved. Some older fossils, such as the upper Triassic *Sanmiguelia*, have been suggested. Based on current evidence, some propose that the ancestors of the angiosperms diverged from an unknown group of gymnosperms in the Triassic period (245–202 million years ago). Fossil angiosperm-like pollen from the Middle Triassic (247.2–242.0 Ma) suggests an older date for their origin. A close relationship between angiosperms and gnetophytes, proposed on the basis of morphological evidence, has more recently been disputed on the basis of molecular evidence that suggest gnetophytes are instead more closely related to other gymnosperms.

The evolution of seed plants and later angiosperms appears to be the result of two distinct rounds of whole genome duplication events. These occurred at 319 million years ago and 192 million years ago. Another possible whole genome duplication event at 160 million years ago perhaps created the ancestral line that led to all modern flowering plants. That event was studied by sequencing the genome of an ancient flowering plant, *Amborella trichopoda*, and directly addresses Darwin's "*abominable mystery*."

The earliest known macrofossil confidently identified as an angiosperm, *Archaefructus liaoningensis*, is dated to about 125 million years BP (the Cretaceous period), whereas pollen considered to be of angiosperm origin takes the fossil record back to about 130 million years BP. However, one study has suggested that the early-middle Jurassic plant *Schmeissneria*, traditionally considered a type of ginkgo, may be the earliest known angiosperm, or at least a close relative. In addition, circumstantial chemical evidence has been found for the existence of angiosperms as early as 250 million years ago. Oleanane, a secondary metabolite produced by many flowering plants, has been found in Permian deposits of that age together with fossils of gigantopterids. Gigantopterids are a group of extinct seed plants that share many morphological traits with flowering plants, although they are not known to have been flowering plants themselves.

In 2013 flowers encased in amber were found and dated 100 million years before present. The amber had frozen the act of sexual reproduction in the process of taking place. Microscopic images showed tubes growing out of pollen and penetrating the flower's stigma. The pollen was sticky, suggesting it was carried by insects.

Recent DNA analysis based on molecular systematics showed that *Amborella trichopoda*, found on the Pacific island of New Caledonia, belongs to a sister group of the other flowering plants, and morphological studies suggest that it has features that may have been characteristic of the earliest flowering plants.

The orders Amborellales, Nymphaeales, and Austrobaileyales diverged as separate lineages from the remaining angiosperm clade at a very early stage in flowering plant evolution. The great angiosperm radiation, when a great diversity of angiosperms appears in the fossil record, occurred in the mid-Cretaceous (approximately 100 million years ago). However, a study in 2007 estimated that the division of the five most recent (the genus *Ceratophyllum*, the family Chloranthaceae, the eudicots, the magnoliids, and the monocots) of the eight main groups occurred around 140 million years ago. By the late Cretaceous, angiosperms appear to have dominated environments formerly occupied by ferns and cycadophytes, but large canopy-forming trees replaced conifers as the dominant trees only close to the end of the Cretaceous 66 million years ago or even later, at the beginning of the Tertiary. The radiation of herbaceous angiosperms occurred much later. Yet, many fossil plants recognizable as belonging to modern families (including beech, oak, maple, and magnolia) had already appeared by the late Cretaceous.

Two bees on a flower head of Creeping Thistle, *Cirsium arvense*

It is generally assumed that the function of flowers, from the start, was to involve mobile animals in their reproduction processes. That is, pollen can be scattered even if the flower is not brightly colored or oddly shaped in a way that attracts animals; however, by expending the energy required to create such traits, angiosperms can enlist the aid of animals and, thus, reproduce more efficiently. Island genetics provides one proposed explanation for the sudden, fully developed appearance of flowering plants. Island genetics is believed to be a common source of speciation in general, especially when it comes to radical adaptations that seem to have required inferior transitional forms. Flowering plants may have evolved in an isolated setting like an island or island chain, where the plants bearing them were able to develop a highly specialized relationship with some specific animal (a wasp, for example). Such a relationship, with a hypothetical wasp carrying pollen from one plant to another much the way fig wasps do today, could result in the development of a high degree of specialization in both the plant(s) and their partners. Note that the wasp example is not incidental; bees, which, it is postulated, evolved specifically due to mutualistic plant relationships, are descended from wasps.

Animals are also involved in the distribution of seeds. Fruit, which is formed by the enlargement of flower parts, is frequently a seed-dispersal tool that attracts animals to eat or otherwise disturb it, incidentally scattering the seeds it contains (see frugivory). Although many such mutualistic relationships remain too fragile to survive competition and to spread widely, flowering proved to be an unusually effective means of reproduction, spreading (whatever its origin) to become the dominant form of land plant life.

Flower ontogeny uses a combination of genes normally responsible for forming new shoots. The most primitive flowers probably had a variable number of flower parts, often separate from (but in contact with) each other. The flowers tended to grow in a spiral pattern, to be bisexual (in plants, this means both male and female parts on the same flower), and to be dominated by the ovary (female part). As flowers evolved, some variations developed parts fused together, with a much more specific number and design, and with either specific sexes per flower or plant or at least "ovary-inferior".

Flower evolution continues to the present day; modern flowers have been so profoundly influenced by humans that some of them cannot be pollinated in nature. Many modern domesticated flower species were formerly simple weeds, which sprouted only when the ground was disturbed. Some of them tended to grow with human crops, perhaps already having symbiotic companion plant relationships with them, and the prettiest did not get plucked because of their beauty, developing a dependence upon and special adaptation to human affection.

A few paleontologists have also proposed that flowering plants, or angiosperms, might have evolved due to interactions with dinosaurs. One of the idea's strongest proponents is Robert T. Bakker. He proposes that herbivorous dinosaurs, with their eating habits, provided a selective pressure on plants, for which adaptations either succeeded in deterring or coping with predation by herbivores.

Classification

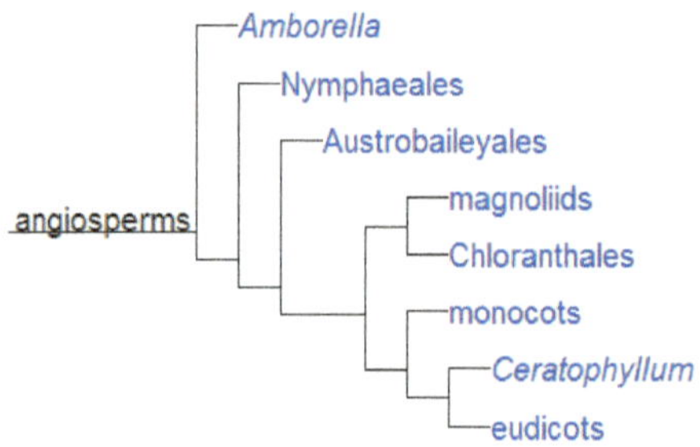

1. Phylogeny of the flowering plants, as of APG III (2009).

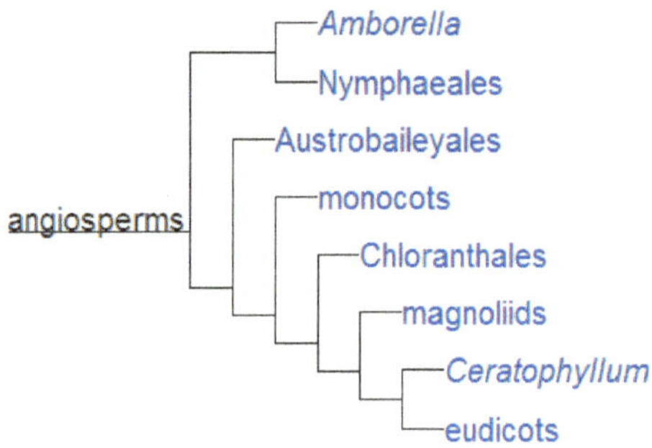

2. Example of alternative phylogeny (2010)

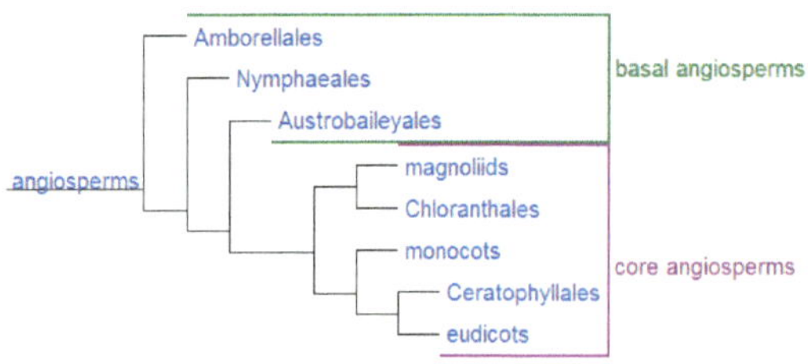

3. APG IV (2016)

There are eight groups of living angiosperms:

- *Amborella*, a single species of shrub from New Caledonia;
- Nymphaeales, about 80 species, water lilies and Hydatellaceae;
- Austrobaileyales, about 100 species of woody plants from various parts of the world;
- Chloranthales, several dozen species of aromatic plants with toothed leaves;
- Magnoliids, about 9,000 species, characterized by trimerous flowers, pollen with one pore, and usually branching-veined leaves—for example magnolias, bay laurel, and black pepper;

- Monocots, about 70,000 species, characterized by trimerous flowers, a single cotyledon, pollen with one pore, and usually parallel-veined leaves—for example grasses, orchids, and palms;
- *Ceratophyllum*, about 6 species of aquatic plants, perhaps most familiar as aquarium plants;
- Eudicots, about 175,000 species, characterized by 4- or 5-merous flowers, pollen with three pores, and usually branching-veined leaves—for example sunflowers, petunia, buttercup, apples, and oaks.

The exact relationship between these eight groups is not yet clear, although there is agreement that the first three groups to diverge from the ancestral angiosperm were Amborellales, Nymphaeales, and Austrobaileyales. The term basal angiosperms refers to these three groups. Among the rest, the relationship between the three broadest of these groups (magnoliids, monocots, and eudicots) remains unclear. Some analyses make the magnoliids the first to diverge, others the monocots. *Ceratophyllum* seems to group with the eudicots rather than with the monocots.

History of Classification

The botanical term "Angiosperm", from the Ancient Greek αγγείον, *angeíon* (bottle, vessel) and σπέρμα, (seed), was coined in the form Angiospermae by Paul Hermann in 1690, as the name of one of his primary divisions of the plant kingdom. This included flowering plants possessing seeds enclosed in capsules, distinguished from his Gymnospermae, or flowering plants with achenial or schizo-carpic fruits, the whole fruit or each of its pieces being here regarded as a seed and naked. The term and its antonym were maintained by Carl Linnaeus with the same sense, but with restricted application, in the names of the orders of his class Didynamia. Its use with any approach to its modern scope became possible only after 1827, when Robert Brown established the existence of truly naked ovules in the Cycadeae and Coniferae, and applied to them the name Gymnosperms. From that time onward, as long as these Gymnosperms were, as was usual, reckoned as dicotyledonous flowering plants, the term Angiosperm was used antithetically by botanical writers, with varying scope, as a group-name for other dicotyledonous plants.

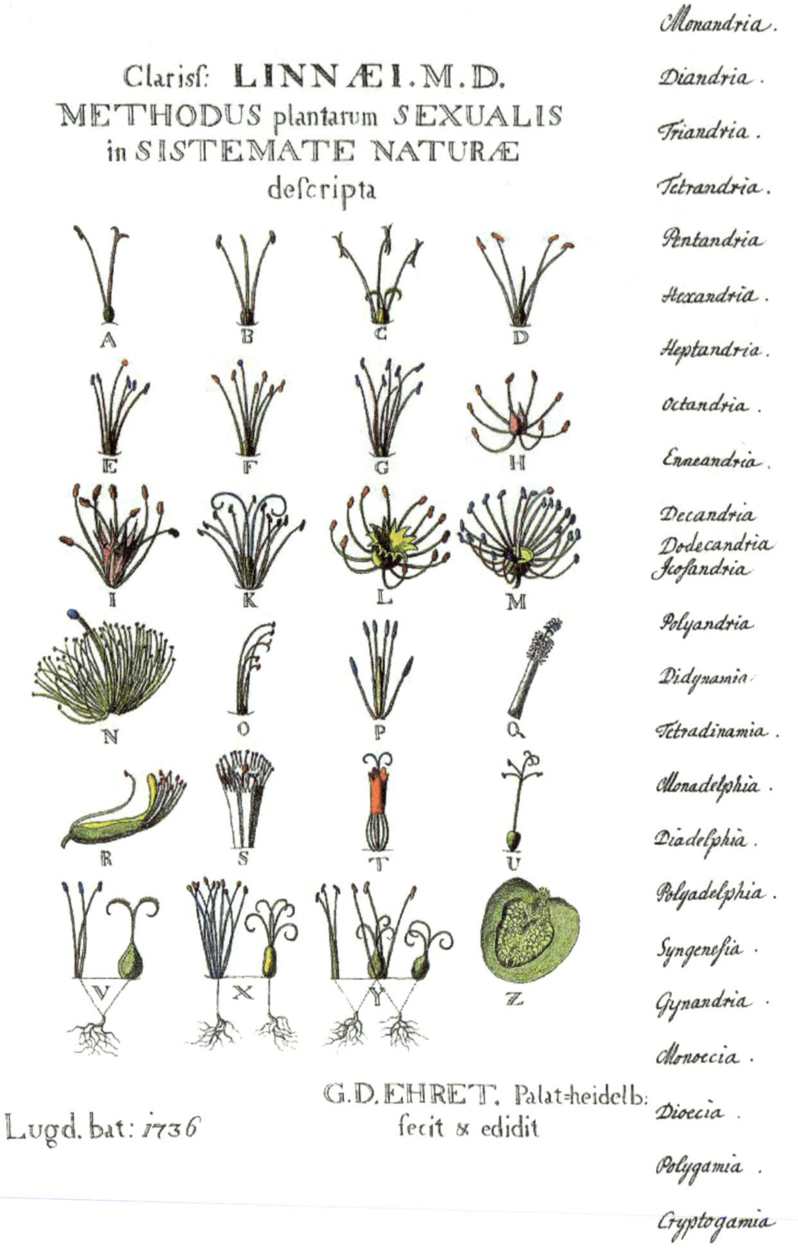

From 1736, an illustration of Linnaean classification.

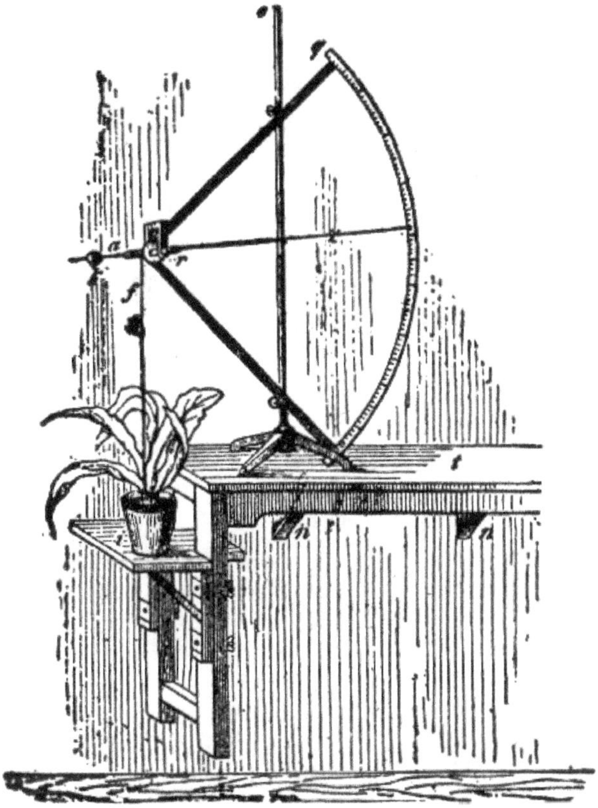

Auxanometer: Device for measuring increase or rate of growth in plants

In 1851, Hofmeister discovered the changes occurring in the embryo-sac of flowering plants, and determined the correct relationships of these to the Cryptogamia. This fixed the position of Gymnosperms as a class distinct from Dicotyledons, and the term Angiosperm then gradually came to be accepted as the suitable designation for the whole of the flowering plants other than Gymnosperms, including the classes of Dicotyledons and Monocotyledons. This is the sense in which the term is used today.

In most taxonomies, the flowering plants are treated as a coherent group. The most popular descriptive name has been Angiospermae (Angiosperms), with Anthophyta ("flowering plants") a second choice. These names are not linked to any rank. The Wettstein system and the Engler system use the name Angiospermae, at the assigned rank of subdivision. The Reveal system treated flowering plants as subdivision Magnoliophytina (Frohne & U. Jensen ex Reveal, Phytologia 79: 70 1996), but later split it to Magnoliopsida, Liliopsida, and Rosopsida. The Takhtajan system and Cronquist system treat this group at the rank of division, leading to the name Magnoliophyta (from the family name Magnoliaceae). The Dahlgren system and Thorne system (1992) treat this group at the rank of class, leading to the name Magnoliopsida. The APG system of 1998, and the later 2003 and 2009 revisions, treat the flowering plants as a clade called angiosperms without a formal botanical name. However, a formal classification was published alongside the 2009 revision in which the flowering plants form the Subclass Magnoliidae.

The internal classification of this group has undergone considerable revision. The Cronquist system, proposed by Arthur Cronquist in 1968 and published in its full form in 1981, is still widely used but is no longer believed to accurately reflect phylogeny. A consensus about how the flowering plants should be arranged has recently begun to emerge through the work of the Angiosperm Phylogeny Group (APG), which published an influential reclassification of the angiosperms in 1998. Updates incorporating more recent research were published as APG II in 2003 and as APG III in 2009.

Monocot (left) and dicot seedlings

Traditionally, the flowering plants are divided into two groups, which in the Cronquist system are called Magnoliopsida (at the rank of class, formed from the family name Magnoliaceae) and Liliopsida (at the rank of class, formed from the family name Liliaceae). Other descriptive names allowed by Article 16 of the ICBN include Dicotyledones or Dicotyledoneae, and Monocotyledones or Monocotyledoneae, which have a long history of use. In English a member of either group may be called a dicotyledon (plural dicotyledons) and monocotyledon (plural monocotyledons), or abbreviated, as dicot (plural dicots) and monocot (plural monocots). These names derive from the observation that the dicots most often have two cotyledons, or embryonic leaves, within each seed. The monocots usually have only one, but the rule is not absolute either way. From a broad diagnostic point of view, the number of cotyledons is neither a particularly handy nor a reliable character.

Recent studies, as by the APG, show that the monocots form a monophyletic group (clade) but that the dicots do not (they are paraphyletic). Nevertheless, the majority of dicot species do form a monophyletic group, called the eudicots or tricolpates. Of the remaining dicot species, most belong to a third major clade known as the magnoliids, containing about 9,000 species. The rest include a paraphyletic grouping of primitive species known collectively as the basal angiosperms, plus the families Ceratophyllaceae and Chloranthaceae.

Flowering Plant Diversity

A poster of twelve different species of flowers of the *Asteraceae* family

Lupinus pilosus

The number of species of flowering plants is estimated to be in the range of 250,000 to 400,000. This compares to around 12,000 species of moss or 11,000 species of pteridophytes, showing that the flowering plants are much more diverse. The number of families in APG (1998) was 462. In APG II (2003) it is not settled; at maximum it is 457, but within this number there are 55 optional segregates, so that the minimum number of families in this system is 402. In APG III (2009) there are 415 families.

The diversity of flowering plants is not evenly distributed. Nearly all species belong to the eudicot (75%), monocot (23%), and magnoliid (2%) clades. The remaining 5 clades contain a little over 250 species in total; i.e. less than 0.1% of flowering plant diversity, divided among 9 families. The 42 most-diverse of 443 families of flowering plants by species, in their APG circumscriptions, are

1. Asteraceae or Compositae (daisy family): 22,750 species;
2. Orchidaceae (orchid family): 21,950;
3. Fabaceae or Leguminosae (bean family): 19,400;
4. Rubiaceae (madder family): 13,150;
5. Poaceae or Gramineae (grass family): 10,035;
6. Lamiaceae or Labiatae (mint family): 7,175;
7. Euphorbiaceae (spurge family): 5,735;
8. Melastomataceae or Melastomaceae (melastome family): 5,005;
9. Myrtaceae (myrtle family): 4,625;
10. Apocynaceae (dogbane family): 4,555;
11. Cyperaceae (sedge family): 4,350;
12. Malvaceae (mallow family): 4,225;
13. Araceae (arum family): 4,025;
14. Ericaceae (heath family): 3,995;
15. Gesneriaceae (gesneriad family): 3,870;
16. Apiaceae or Umbelliferae (parsley family): 3,780;
17. Brassicaceae or Cruciferae (cabbage family): 3,710:
18. Piperaceae (pepper family): 3,600;
19. Acanthaceae (acanthus family): 3,500;

20. Rosaceae (rose family): 2,830;
21. Boraginaceae (borage family): 2,740;
22. Urticaceae (nettle family): 2,625;
23. Ranunculaceae (buttercup family): 2,525;
24. Lauraceae (laurel family): 2,500;
25. Solanaceae (nightshade family): 2,460;
26. Campanulaceae (bellflower family): 2,380;
27. Arecaceae (palm family): 2,361;
28. Annonaceae (custard apple family): 2,220;
29. Caryophyllaceae (pink family): 2,200;
30. Orobanchaceae (broomrape family): 2,060;
31. Amaranthaceae (amaranth family): 2,050;
32. Iridaceae (iris family): 2,025;
33. Aizoaceae or Ficoidaceae (ice plant family): 2,020;
34. Rutaceae (rue family): 1,815;
35. Phyllanthaceae (phyllanthus family): 1,745;
36. Scrophulariaceae (figwort family): 1,700;
37. Gentianaceae (gentian family): 1,650;
38. Convolvulaceae (bindweed family): 1,600;
39. Proteaceae (protea family): 1,600;
40. Sapindaceae (soapberry family): 1,580;
41. Cactaceae (cactus family): 1,500;
42. Araliaceae (*Aralia* or ivy family): 1,450.

Of these, the Orchidaceae, Poaceae, Cyperaceae, Arecaceae, and Iridaceae are monocot families; Piperaceae, Lauraceae, and Annonaceae are magnoliid dicots; the rest of the families are eudicots.

Vascular Anatomy

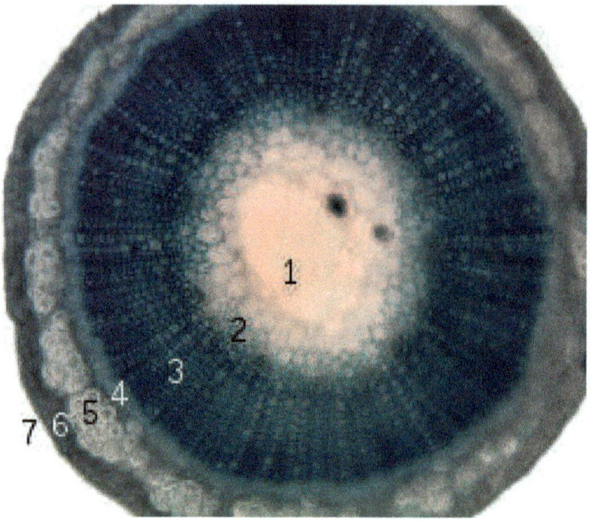

Cross-section of a stem of the angiosperm flax:

- Pith,
- Protoxylem,
- Xylem I,
- Phloem I,
- Sclerenchyma (bast fibre),
- Cortex,
- Epidermis

The amount and complexity of tissue-formation in flowering plants exceeds that of gymnosperms. The vascular bundles of the stem are arranged such that the xylem and phloem form concentric rings.

In the dicotyledons, the bundles in the very young stem are arranged in an open ring, separating a central pith from an outer cortex. In each bundle, separating the xylem and phloem, is a layer of meristem or active formative tissue known as cambium. By the formation of a layer of cambium between the bundles (interfascicular cambium), a complete ring is formed, and a regular periodical increase in thickness results from the development of xylem on the inside and phloem on the outside. The soft phloem becomes crushed, but the hard wood persists and forms the bulk of the stem and branches of the woody perennial. Owing to differences in the character of the elements produced at the beginning and end of the season, the wood is marked out in transverse section into concentric rings, one for each season of growth, called annual rings.

Among the monocotyledons, the bundles are more numerous in the young stem and are scattered through the ground tissue. They contain no cambium and once formed the stem increases in diameter only in exceptional cases.

The Flower, Fruit, and Seed

Flowers

A collection of flowers forming an inflorescence

The characteristic feature of angiosperms is the flower. Flowers show remarkable variation in form and elaboration, and provide the most trustworthy external characteristics for establishing relationships among angiosperm species. The function of the flower is to ensure fertilization of the ovule and development of fruit containing seeds. The floral apparatus may arise terminally on a shoot or from the axil of a leaf (where the petiole attaches to the stem). Occasionally, as in violets, a flower arises singly in the axil of an ordinary foliage-leaf. More typically, the flower-bearing portion of the plant is sharply distinguished from the foliage-bearing or vegetative portion, and forms a more or less elaborate branch-system called an inflorescence.

There are two kinds of reproductive cells produced by flowers. Microspores, which will divide to become pollen grains, are the "male" cells and are borne in the stamens (or microsporophylls). The "female" cells called megaspores, which will divide to become the egg cell (megagametogenesis), are contained in the ovule and enclosed in the carpel (or megasporophyll).

The flower may consist only of these parts, as in willow, where each flower comprises only a few stamens or two carpels. Usually, other structures are present and serve to protect the sporophylls and to form an envelope attractive to pollinators. The individual members of these surrounding structures are known as sepals and petals (or tepals in flowers such as *Magnolia* where sepals and petals are not distinguishable from each other). The outer series (calyx of sepals) is usually green and leaf-like, and functions to protect the rest of the flower, especially the bud. The inner series (corolla of petals) is, in general, white or brightly colored, and is more delicate in structure. It functions to attract insect or bird pollinators. Attraction is effected by color, scent, and nectar, which may be secreted in some part of the flower. The characteristics that attract pollinators account for the popularity of flowers and flowering plants among humans.

While the majority of flowers are perfect or hermaphrodite (having both pollen and ovule producing parts in the same flower structure), flowering plants have developed numerous morphological and physiological mechanisms to reduce or prevent self-fertilization. Heteromorphic flowers have short carpels and long stamens, or vice versa, so animal pollinators cannot easily transfer pollen to the pistil (receptive part of the carpel). Homomorphic flowers may employ a biochemical (physiological) mechanism called self-incompatibility to discriminate between self and non-self pollen grains. In other species, the male and female parts are morphologically separated, developing on different flowers.

Fertilization and Embryogenesis

Double fertilization refers to a process in which two sperm cells fertilize cells in the ovary. This process begins when a pollen grain adheres to the stigma of the pistil (female reproductive structure), germinates, and grows a long pollen tube. While this pollen tube is growing, a haploid generative cell travels down the tube behind the tube nucleus. The generative cell divides by mitosis to produce two haploid (n) sperm cells. As the pollen tube grows, it makes its way from the stigma, down the style and into the ovary. Here the pollen tube reaches the micropyle of the ovule and digests its way into one of the synergids, releasing its contents (which include the sperm cells). The synergid that the cells were released into degenerates and one sperm makes its way to fertilize the egg cell, producing a diploid ($2n$) zygote. The second sperm cell fuses with both central cell nuclei, producing a triploid ($3n$) cell. As the zygote develops into an embryo, the triploid cell develops into the endosperm, which serves as the embryo's food supply. The ovary will now develop into a fruit and the ovule will develop into a seed.

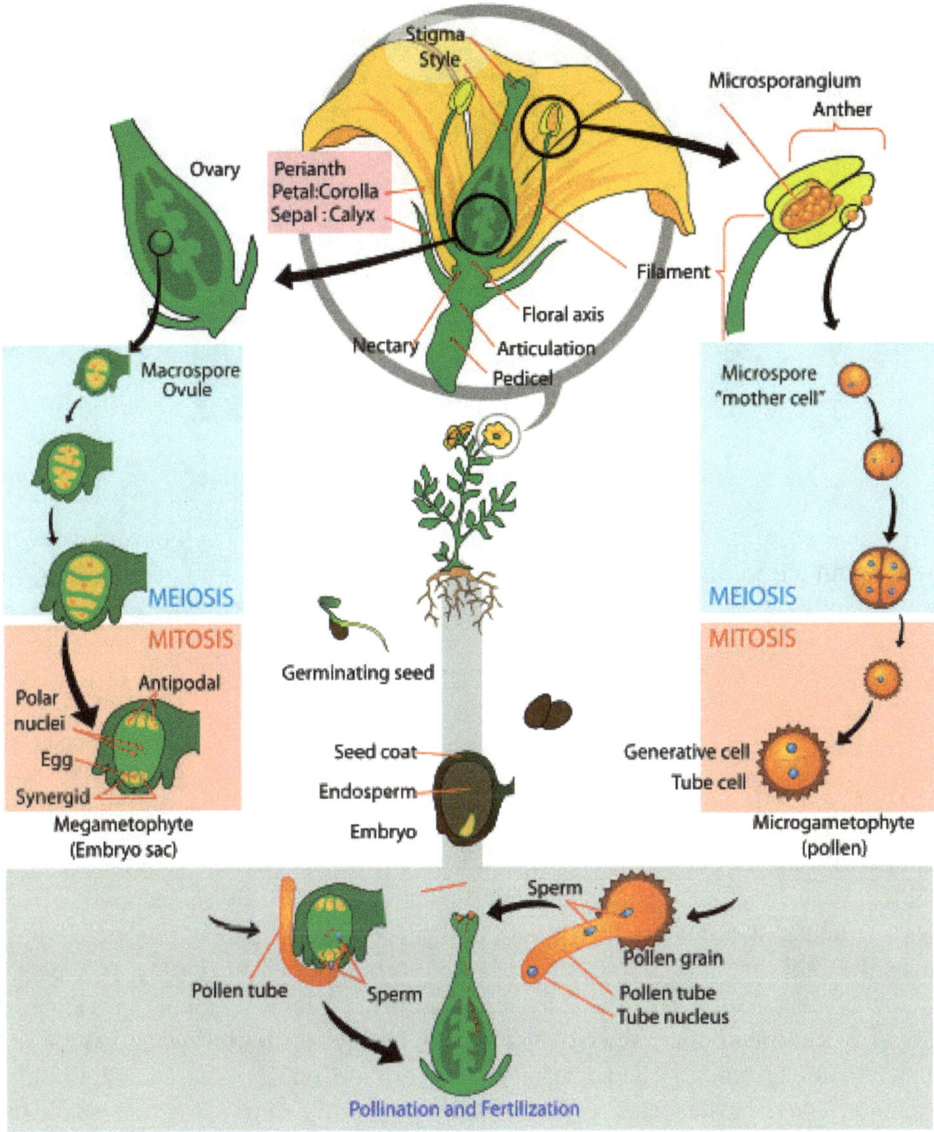

Angiosperm life cycle

Fruit and Seed

As the development of embryo and endosperm proceeds within the embryo sac, the sac wall enlarges and combines with the nucellus (which is likewise enlarging) and the integument to form the *seed coat*. The ovary wall develops to form the fruit or pericarp, whose form is closely associated with the manner of distribution of the seed.

Frequently, the influence of fertilization is felt beyond the ovary, and other parts of the flower take part in the formation of the fruit, e.g., the floral receptacle in the apple, strawberry, and others.

The fruit of the *Aesculus* or Horse Chestnut tree

The character of the seed coat bears a definite relation to that of the fruit. They protect the embryo and aid in dissemination; they may also directly promote germination. Among plants with indehiscent fruits, in general, the fruit provides protection for the embryo and secures dissemination. In this case, the seed coat is only slightly developed. If the fruit is dehiscent and the seed is exposed, in general, the seed-coat is well developed, and must discharge the functions otherwise executed by the fruit.

Economic Importance

Agriculture is almost entirely dependent on angiosperms, which provide virtually all plant-based food, and also provide a significant amount of livestock feed. Of all the families of plants, the Poaceae, or grass family (grains), is by far the most important, providing the bulk of all feedstocks (rice, corn — maize, wheat, barley, rye, oats, pearl millet, sugar cane, sorghum). The Fabaceae, or legume family, comes in second place. Also of high importance are the Solanaceae, or nightshade family (potatoes, tomatoes, and peppers, among others), the Cucurbitaceae, or gourd family (also including pumpkins and melons), the Brassicaceae, or mustard plant family (including rapeseed and the innumerable varieties of the cabbage species *Brassica oleracea*), and the Apiaceae, or parsley family. Many of our fruits come from the Rutaceae, or rue family (including oranges, lemons, grapefruits, etc.), and the Rosaceae, or rose family (including apples, pears, cherries, apricots, plums, etc.).

In some parts of the world, certain single species assume paramount importance because of their variety of uses, for example the coconut (*Cocos nucifera*) on Pacific atolls, and the olive (*Olea europaea*) in the Mediterranean region.

Flowering plants also provide economic resources in the form of wood, paper, fiber (cotton, flax, and hemp, among others), medicines (digitalis, camphor), decorative and landscaping plants, and many other uses. The main area in which they are surpassed by other plants — namely, coniferous trees (Pinales), which are non-flowering (gymnosperms) — is timber and paper production.

REFERENCES

1. Cavalier-Smith, T. (1981). "Eukaryote kingdoms: Seven or nine?". BioSystems. 14 (3–4): 461–481. doi:10.1016/0303-2647(81)90050-2.
2. Lewis, L.A.; McCourt, R.M. (2004). "Green algae and the origin of land plants". American Journal of Botany. 91: 1535–1556. doi:10.3732/ajb.91.10.1535.
3. Kenrick, Paul; Crane, Peter R. (1997). The origin and early diversification of land plants: A cladistic study. Washington, D. C.: Smithsonian Institution Press. ISBN 1-56098-730-8.
4. Adl, S.M. et al. (2005). "The new higher level classification of eukaryotes with emphasis on the taxonomy of protists". Journal of Eukaryote Microbiology. 52: 399–451. doi:10.1111/j.1550-7408.2005.00053.x.
5. "Numbers of threatened species by major groups of organisms (1996–2010)" (PDF). International Union for Conservation of Nature. 11 Mar 2010.
6. Field, C.B.; Behrenfeld, M.J.; Randerson, J.T.; Falkowski, P. (1998). "Primary production of the biosphere: Integrating terrestrial and oceanic components". Science. 281 (5374): 237–240. Bibcode:1998Sci...281..237F. doi:10.1126/science.281.5374.237. PMID 9657713.
7. "plant - Definition from the Merriam-Webster Online Dictionary". Retrieved 2009-03-25.
8. "plant (life form) -- Britannica Online Encyclopedia". Retrieved 2009-03-25.
9. Whittaker, R. H. (1969). "New concepts of kingdoms or organisms" (PDF). Science. 163 (3863): 150–160. doi:10.1126/science.163.3863.150. PMID 5762760.
10. Margulis, L (1971). "Whittaker's five kingdoms of organisms: minor revisions suggested by considerations of the origin of mitosis". Evolution. 25: 242–245. doi:10.2307/2406516.
11. Copeland, H. F. (1956). The Classification of Lower Organisms. Palo Alto: Pacific Books, p. 6, .
12. Cavalier-Smith, T. (1981). "Eukaryote Kingdoms: Seven or Nine?".". BioSystems. 14 (3–4): 461–481. doi:10.1016/0303-2647(81)90050-2. PMID 7337818.
13. Linnaeus, C. (1751). Philosophia botanica, 1st ed., p. 37.
14. Haeckel, E. (1866). Generale Morphologie der Organismen. Berlin: Verlag von Georg Reimer. pp. vol.1: i–xxxii, 1–574, pls I–II; vol. 2: i–clx, 1–462, pls I–VIII.
15. Haeckel, E. (1894). Die systematische Phylogenie.
16. Based on Rogozin, I.B.; Basu, M.K.; Csürös, M. & Koonin, E.V. (2009), "Analysis of Rare Genomic Changes Does Not Support the Unikont–Bikont Phylogeny and Suggests Cyanobacterial Symbiosis as the Point of Primary Radiation of Eukaryotes", Genome Biology and Evolution, 1: 99–113, doi:10.1093/gbe/evp011, PMC 2817406Freely accessible, PMID 20333181 and Becker, B. & Marin, B. (2009), "Streptophyte algae and the origin of embryophytes", Annals of Botany, 103 (7): 999–1004, doi:10.1093/aob/mcp044, PMC 2707909Freely accessible, PMID 19273476; see also the slightly different cladogram in Lewis, Louise A.

& McCourt, R.M. (2004), "Green algae and the origin of land plants", Am. J. Bot., 91 (10): 1535–1556, doi:10.3732/ajb.91.10.1535, PMID 21652308

17. Parfrey, Laura Wegener; Lahr, Daniel J. G.; Knoll, Andrew H.; Katz, Laura A. (2011-08-16). "Estimating the timing of early eukaryotic diversification with multigene molecular clocks". Proceedings of the National Academy of Sciences. 108 (33): 13624–13629. doi:10.1073/pnas.1110633108. ISSN 0027-8424. PMC 3158185Freely accessible. PMID 21810989.
18. Derelle, Romain; Torruella, Guifré; Klimeš, Vladimír; Brinkmann, Henner; Kim, Eunsoo; Vlček, Čestmír; Lang, B. Franz; Eliáš, Marek (2015-02-17). "Bacterial proteins pinpoint a single eukaryotic root". Proceedings of the National Academy of Sciences. 112 (7): E693–E699. doi:10.1073/pnas.1420657112. ISSN 0027-8424. PMC 4343179Freely accessible. PMID 25646484.
19. Jackson, Christopher; Clayden, Susan; Reyes-Prieto, Adrian (2015-01-01). "The Glaucophyta: the blue-green plants in a nutshell". Acta Societatis Botanicorum Poloniae. 84 (2): 149–165. doi:10.5586/asbp.2015.020.
20. Margulis, L. (1974). "Five-kingdom classification and the origin and evolution of cells". Evolutionary Biology. 7: 45–78. doi:10.1007/978-1-4615-6944-2_2.
21. Raven, Peter H.; Evert, Ray F.; Eichhorn, Susan E. (2005). Biology of Plants (7th ed.). New York: W. H. Freeman and Company. ISBN 0-7167-1007-2.
22. Lewis, Louise A. & McCourt, R.M. (2004), "Green algae and the origin of land plants", Am. J. Bot., 91 (10): 1535–1556, doi:10.3732/ajb.91.10.1535, PMID 21652308
23. Becker, B. & Marin, B. (2009), "Streptophyte algae and the origin of embryophytes", Annals of Botany, 103 (7): 999–1004, doi:10.1093/aob/mcp044, PMC 2707909Freely accessible, PMID 19273476
24. Guiry, M.D. & Guiry, G.M. (2007). "Phylum: Chlorophyta taxonomy browser". AlgaeBase version 4.2 World-wide electronic publication, National University of Ireland, Galway. Retrieved 2007-09-23.
25. Deacon, J.W. (2005). Fungal Biology. Wiley. ISBN 9781405130660.
26. Van den Hoek, C., D. G. Mann, & H. M. Jahns, 1995. Algae: An Introduction to Phycology. pages 343, 350, 392, 413, 425, 439, & 448 (Cambridge: Cambridge University Press). ISBN 0-521-30419-9
27. Guiry, M.D. & Guiry, G.M. (2011), AlgaeBase : Chlorophyta, World-wide electronic publication, National University of Ireland, Galway, retrieved 2011-07-26
28. Guiry, M.D. & Guiry, G.M. (2011), AlgaeBase : Charophyta, World-wide electronic publication, National University of Ireland, Galway, retrieved 2011-07-26
29. Van den Hoek, C., D. G. Mann, & H. M. Jahns, 1995. Algae: An Introduction to Phycology. pages 457, 463, & 476. (Cambridge: Cambridge University Press). ISBN 0-521-30419-9
30. Crandall-Stotler, Barbara. & Stotler, Raymond E., 2000. "Morphology and

classification of the Marchantiophyta". page 21 in A. Jonathan Shaw & Bernard Goffinet (Eds.), Bryophyte Biology. (Cambridge: Cambridge University Press). ISBN 0-521-66097-1

31. Schuster, Rudolf M., The Hepaticae and Anthocerotae of North America, volume VI, pages 712–713. (Chicago: Field Museum of Natural History, 1992). ISBN 0-914868-21-7.
32. Goffinet, Bernard; William R. Buck (2004). "Systematics of the Bryophyta (Mosses): From molecules to a revised classification". Monographs in Systematic Botany. Missouri Botanical Garden Press. 98: 205–239.
33. Gifford, Ernest M.; Foster, Adriance S. (1988). Morphology and Evolution of Vascular Plants (3rd ed.). New York: W. H. Freeman and Company. p. 358. ISBN 0-7167-1946-0.
34. Taylor, Thomas N.; Taylor, Edith L. (1993). The Biology and Evolution of Fossil Plants. New Jersey: Prentice-Hall. p. 636. ISBN 0-13-651589-4.
35. International Union for Conservation of Nature and Natural Resources, 2006. IUCN Red List of Threatened Species:Summary Statistics
36. "The oldest fossils reveal evolution of non-vascular plants by the middle to late Ordovician Period (≈450–440 m.y.a.) on the basis of fossil spores" Transition of plants to land
37. "The apparent dominance of eukaryotes in non-marine settings by 1 Gyr ago indicates that eukaryotic evolution on land may have commenced far earlier than previously thought." Earth's earliest non-marine eukaryotes
38. Plants crawled onto land earlier than we give them credit, genetic evidence suggests
39. Rothwell, G. W.; Scheckler, S. E.; Gillespie, W. H. (1989). "Elkinsia gen. nov., a Late Devonian gymnosperm with cupulate ovules". Botanical Gazette. University of Chicago Press. 150 (2): 170–189. doi:10.1086/337763. JSTOR 2995234.
40. Kenrick, Paul & Peter R. Crane. 1997. The Origin and Early Diversification of Land Plants: A Cladistic Study. (Washington, D.C., Smithsonian Institution Press.) ISBN 1-56098-730-8.
41. Smith Alan R.; Pryer Kathleen M.; Schuettpelz E.; Korall P.; Schneider H.; Wolf Paul G. (2006). "A classification for extant ferns" (PDF). Taxon. 55 (3): 705–731. doi:10.2307/25065646.
42. Leliaert, F., Smith, D.R., Moreau, H., Herron, M.D., Verbruggen, H., Delwiche, C.F. & De Clerck, O. (2012). "Phylogeny and molecular evolution of the green algae" (PDF). Critical Reviews in Plant Sciences. 31: 1–46. doi:10.1080/07352689.2011.615705.
43. Leliaert, Frederik; Verbruggen, Heroen; Zechman, Frederick W. (2011). "Into the deep: New discoveries at the base of the green plant phylogeny". BioEssays. 33 (9): 683–692. doi:10.1002/bies.201100035. ISSN 0265-9247. PMID 21744372.
44. Silar, Philippe (2016), "Protistes Eucaryotes: Origine, Evolution et Biologie des Microbes Eucaryotes", HAL archives-ouvertes: 1–462

45. Novíkov & Barabaš-Krasni (2015). "Modern plant systematics". Liga-Pres: 685. doi:10.13140/RG.2.1.4745.6164. ISBN 978-966-397-276-3.
46. Knoll, Andrew H (2003). Life on a Young Planet: The First Three Billion Years of Evolution on Earth. Princeton University Press.
47. Tappan, H (1980). Palaeobiology of Plant Protists. Freeman, San Francisco.
48. Butterfield, Nicholas J.; Knoll, Andrew H.; Swett, Keene (1994). "Paleobiology of the Neoproterozoic Svanbergfjellet Formation, Spitsbergen". Lethaia. 27 (1): 76–76. doi:10.1111/j.1502-3931.1994.tb01558.x. ISSN 0024-1164.
49. Edward O. Wilson; et al. (1973). Life on Earth (First ed.). p. 145. ISBN 0-87893-934-2.
50. Robbins, W.W., Weier, T.E., et al., Botany:Plant Science, 3rd edition , Wiley International, New York, 1965.
51. Goyal, K., Walton, L. J., & Tunnacliffe, A. (2005). "LEA proteins prevent protein aggregation due to water stress". Biochemical Journal. 388 (Part 1): 151–157. doi:10.1042/BJ20041931. PMC 1186703Freely accessible. PMID 15631617. Archived from the original on 2009-08-03.
52. Glerum, C. 1985. Frost hardiness of coniferous seedlings: principles and applications. p. 107–123 in Duryea, M.L. (Ed.). Proceedings: Evaluating seedling quality: principles, procedures, and predictive abilities of major tests. Workshop, October 1984, Oregon State Univ., For. Res. Lab., Corvallis OR.
53. Lyons, J.M.; Raison, J.K.; Steponkus, P.L. 1979. The plant membrane in response to low temperature: an overview. p. 1–24 in Lyons, J.M.; Graham, D.; Raison, J.K. (Eds.). Low Temperature Stress in Crop Plants. Academic Press, New York NY.
54. Mazur, P. 1977. The role of intracellular freezing in the death of cells cooled at supraoptimal rates. Cryobiology 14:251–272.
55. Sakai, A.; Larcher, W. (Eds.) 1987. Frost Survival of Plants. Springer-Verlag, New York NY. 321 p.
56. Campbell, Reece, Biology, 7th edition, Pearson/Benjamin Cummings, 2005.
57. BBC Report
58. Song, W.Y.; et al. (1995). "A receptor kinase-like protein encoded by the rice disease resistance gene, XA21". Science. 270 (5243): 1804–1806. doi:10.1126/science.270.5243.1804. PMID 8525370.
59. Gomez-Gomez, L.; et al. (2000). "FLS2: an LRR receptor-like kinase involved in the perception of the bacterial elicitor flagellin in Arabidopsis". Molecular Cell. 5 (6): 1003–1011. doi:10.1016/S1097-2765(00)80265-8. PMID 10911994.
60. Howard Frank, Bromeliad Phytotelmata, October 2000
61. Barthlott, W., S. Porembski, R. Seine, and I. Theisen. 2007. The Curious World of Carnivorous Plants: A Comprehensive Guide to Their Biology and Cultivation. Timber Press: Portland, Oregon.
62. "cocaine/crack".

63. "Deaths related to cocaine".
64. "Illegal drugs drain $160 billion a year from American economy". Archived from the original on 2008-02-15.
65. "The social cost of illegal drug consumption in Spain".

CHAPTER 3

Plant Morphology

Chapter Contents

- **Introduction**
- **Scope**
- **A Comparative Science**
- **Cell Wall**
- **Tissues**

The study of the structure and form development of plants in their individual and evolutionary-historical development. It is one of the most important branches of botany. As plant morphology developed, the followingindependent sciences were established: plant anatomy, which studies the tissue and cell structure of organs; plantembryology, which studies embryo development; and cytology, which studies cell structure and development. Thus, plantmorphology, in the narrow sense, primarily studies the form development and structure of an organism; however, it also isconcerned with regularities on the population-species level, inasmuch as it deals with the evolution of form.

INTRODUCTION

Plant morphology or phytomorphology is the study of the physical form and external structure of plants. This is usually considered distinct from plant anatomy, which is the study of the internal structure of plants, especially at the microscopic level. Plant morphology is useful in the visual identification of plants.

Inflorescences emerging from protective coverings

SCOPE

Plant morphology "represents a study of the development, form, and structure of plants, and, by implication, an attempt to interpret these on the basis of similarity of plan and origin." There are four major areas of investigation in plant morphology, and each overlaps with another field of the biological sciences.

Asclepias syriaca showing complex morphology of the flowers.

Looking up into the branch structure of a Pinus sylvestris tree

First of all, morphology is **comparative**, meaning that the morphologist examines structures in many different plants of the same or different species, then draws comparisons and formulates ideas about similarities. When structures in different species are believed to exist and develop as a result of common, inherited genetic pathways, those structures are termed homologous. For example, the leaves of pine, oak, and cabbage all look very different, but share certain basic structures and arrangement of parts. The homology of leaves is an easy conclusion to make. The plant morphologist goes further, and discovers that the spines of cactus also share the same basic structure and development as leaves in other plants, and therefore cactus spines are homologous to leaves as well. This aspect of plant morphology overlaps with the study of plant evolution and paleobotany.

Secondly, plant morphology observes both the vegetative (somatic) structures of plants, as well as the reproductive structures. The vegetative structures of vascular plants includes the study of the shoot system, composed of stems and leaves, as well as the root system. The reproductive structures are more varied, and are usually specific to a particular group of plants, such as flowers and seeds, fern sori, and moss capsules. The detailed study of reproductive structures in plants led to the discovery of the alternation of generations found in all plants and most algae. This area of plant morphology overlaps with the study of biodiversity and plant systematics.

Thirdly, plant morphology studies plant structure at a range of scales. At the smallest scales are ultrastructure, the general structural features of cells visible only with the aid of an electron microscope, and cytology, the study of cells using optical microscopy. At this scale, plant morphology overlaps with plant anatomy as a field of study. At the largest scale is the study of plant growth habit, the overall architecture of a plant. The pattern of branching in a tree will vary from species to species, as will the appearance of a plant as a tree, herb, or grass.

Fourthly, plant morphology examines the pattern of development, the process by which structures originate and mature as a plant grows. While animals produce all the body parts they will ever have from early in their life, plants constantly produce new tissues and structures throughout their life. A living plant always has embryonic tissues. The way in which new structures mature as they are produced may be affected by the point in the plant's life when they begin to develop, as well as by the environment to which the structures are exposed. A morphologist studies this process, the causes, and its result. This area of plant morphology overlaps with plant physiology and ecology.

A COMPARATIVE SCIENCE

A plant morphologist makes comparisons between structures in many different plants of the same or different species. Making such comparisons between similar structures in different plants tackles the question of why the structures are similar. It is quite likely that similar underlying causes of genetics, physiology, or response to the environment have led to this similarity in appearance. The result of scientific investigation into these causes can lead to one of two insights into the underlying biology:

- **Homology** - the structure is similar between the two species because of shared ancestry and common genetics.
- **Convergence** - the structure is similar between the two species because of independent adaptation to common environmental pressures.

Understanding which characteristics and structures belong to each type is an important part of understanding plant evolution. The evolutionary biologist relies on the plant morphologist to interpret structures, and in turn provides phylogenies of plant relationships that may lead to new morphological insights.

Homology

When structures in different species are believed to exist and develop as a result of common, inherited genetic pathways, those structures are termed homologous. For example, the leaves of pine, oak, and cabbage all look very different, but share certain basic structures and arrangement of parts. The homology of leaves is an easy conclusion to make. The plant morphologist goes further, and discovers that the spines of cactus also share the same basic structure and development as leaves in other plants, and therefore cactus spines are homologous to leaves as well.

Convergence

When structures in different species are believed to exist and develop as a result of common adaptive responses to environmental pressure, those structures are termed convergent. For example, the fronds of Bryopsis plumosa and stems of Asparagus setaceus both have the same feathery branching appearance, even though one is an alga and one is a flowering plant. The similarity in overall structure occurs independently as a result of convergence. The growth form of many cacti and species of Euphorbia is very similar, even though they belong to widely distant families. The similarity results from common solutions to the problem of surviving in a hot, dry environment.

Euphorbia obesa, a spurge

Astrophytum asterias, a cactus.

PLANT CELL

Plant cells are eukaryotic cells that differ in several key aspects from the cells of other eukaryotic organisms. Their distinctive features include:

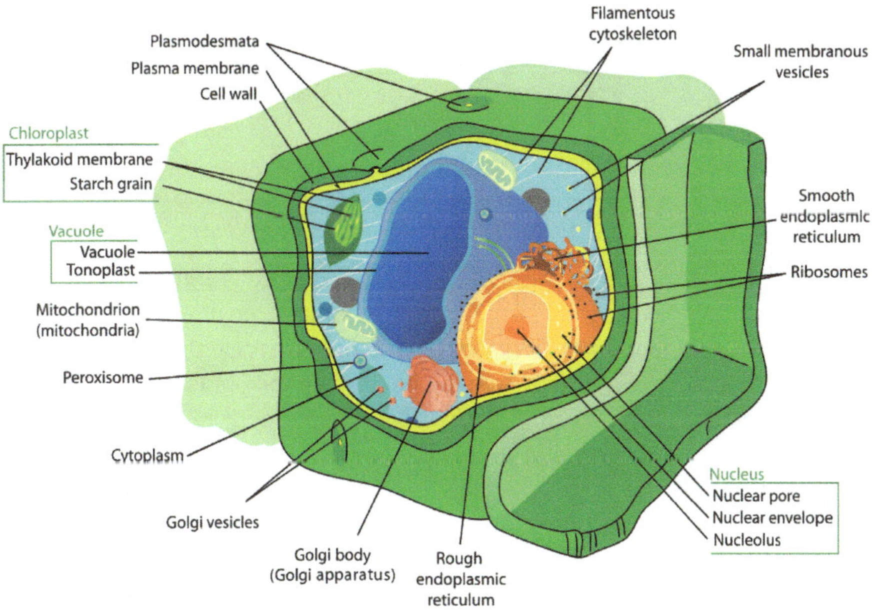

Plant cell structure

- A large central vacuole, a water-filled volume enclosed by a membrane known as the tonoplast that maintains the cell's turgor, controls movement of molecules between the cytosol and sap, stores useful material and digests waste proteins and organelles.
- A cell wall composed of cellulose and hemicellulose, pectin and in many cases lignin, is secreted by the protoplast on the outside of the cell membrane. This contrasts with the cell walls of fungi, which are made of chitin, and of bacteria, which are made of peptidoglycan. Cell walls perform many essential functions:

they provide shape to form the tissue and organs of the plant, and play an important role in intercellular communication and plant-microbe interactions

- Specialized cell-to-cell communication pathways known as plasmodesmata, pores in the primary cell wall through which the plasmalemma and endoplasmic reticulum of adjacent cells are continuous.
- Plastids, the most notable being the chloroplast, which contains chlorophyll, a green-colored pigment that absorbs sunlight, and allows the plant to make its own food in the process known as photosynthesis. Other types of plastids are the amyloplasts, specialized for starch storage, elaioplasts specialized for fat storage, and chromoplasts specialized for synthesis and storage of pigments. As in mitochondria, which have a genome encoding 37 genes, plastids have their own genomes of about 100–120 unique genes and, it is presumed, arose as prokaryotic endosymbionts living in the cells of an early eukaryotic ancestor of the land plants and algae.
- Cell division by construction of a phragmoplast as a template for building a cell plate late in cytokinesis is characteristic of land plants and a few groups of algae, notably the Charophytes and the Order Trentepohliales
- The sperm of bryophytes and pteridophytes, cycads and Ginkgo have flagella similar to those in animal cells, but conifers and flowering plants lack flagella and centrioles.

Cell Types

- Parenchyma cells are living cells that have functions ranging from storage and support to photosynthesis and phloem loading (transfer cells). Apart from the xylem and phloem in their vascular bundles, leaves are composed mainly of parenchyma cells. Some parenchyma cells, as in the epidermis, are specialized for light penetration and focusing or regulation of gas exchange, but others are among the least specialized cells in plant tissue, and may remain totipotent, capable of dividing to produce new populations of undifferentiated cells, throughout their lives. Parenchyma cells have thin, permeable primary walls enabling the transport of small molecules between them, and their cytoplasm is responsible for a wide range of biochemical functions such as nectar secretion, or the manufacture of secondary products that discourage herbivory. Parenchyma cells that contain many chloroplasts and are concerned primarily with photosynthesis are called chlorenchyma cells. Others, such as the majority of the parenchyma cells in potato tubers and the seed cotyledons of legumes, have a storage function.
- Collenchyma cells – collenchyma cells are alive at maturity and have only a primary wall. These cells mature from meristem derivatives that initially resemble parenchyma, but differences quickly become apparent. Plastids do not develop, and the secretory apparatus (ER and Golgi) proliferates to secrete additional primary wall. The wall is most commonly thickest at the corners, where three or more cells come in contact, and thinnest where only two cells come in contact, though other arrangements of the wall thickening are possible.

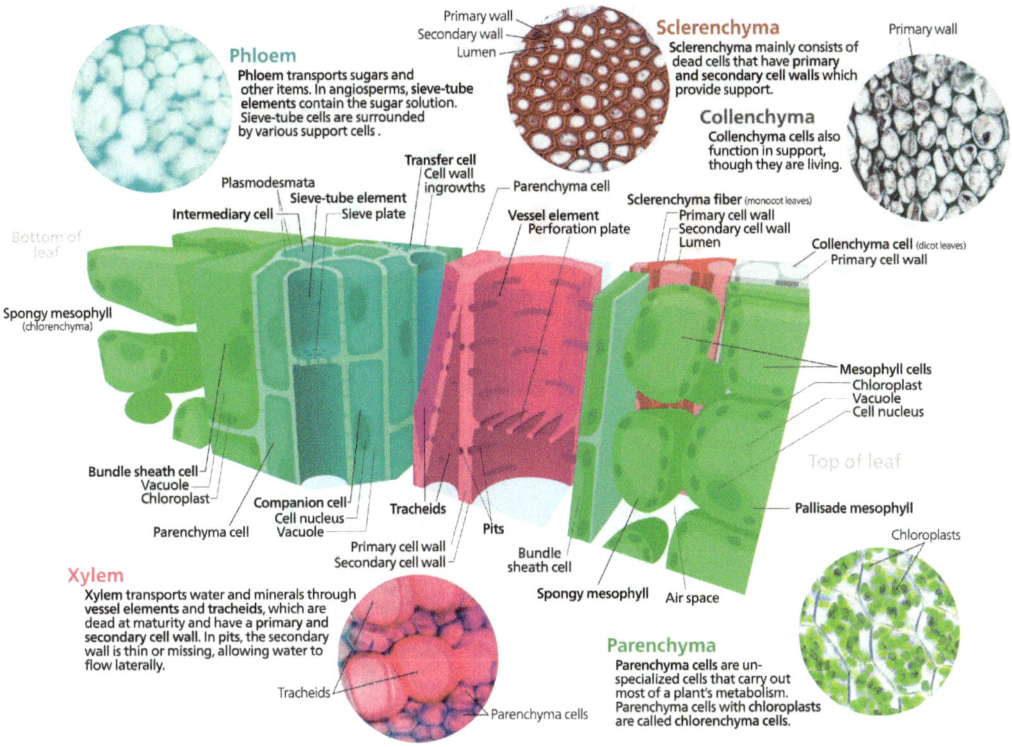

Cross section of a leaf showing various plant cell types

Pectin and hemicellulose are the dominant constituents of collenchyma cell walls of dicotyledon angiosperms, which may contain as little as 20% of cellulose in Petasites. Collenchyma cells are typically quite elongated, and may divide transversely to give a septate appearance. The role of this cell type is to support the plant in axes still growing in length, and to confer flexibility and tensile strength on tissues. The primary wall lacks lignin that would make it tough and rigid, so this cell type provides what could be called plastic support – support that can hold a young stem or petiole into the air, but in cells that can be stretched as the cells around them elongate. Stretchable support (without elastic snap-back) is a good way to describe what collenchyma does. Parts of the strings in celery are collenchyma.

- Sclerenchyma cells – Sclerenchyma cells (from the Greek **skleros**, hard) are hard and tough cells with a function in mechanical support. They are of two broad types – sclereids or stone cells and fibres. The cells develop an extensive secondary cell wall that is laid down on the inside of the primary cell wall. The secondary wall is impregnated with lignin, making it hard and impermeable to water. Thus, these cells cannot survive for long' as they cannot exchange sufficient material to maintain active metabolism. Sclerenchyma cells are typically dead at functional maturity, and the cytoplasm is missing, leaving an empty central cavity.

Functions for sclereid cells (hard cells that give leaves or fruits a gritty texture) include discouraging herbivory, by damaging digestive passages in small insect larval stages, and physical protection (a solid tissue of hard sclereid cells form the pit wall in a peach and many other fruits). Functions of fibres include provision of load-bearing support and tensile strength to the leaves and stems of herbaceous plants. Sclerenchyma fibres are not involved in conduction, either of water and nutrients (as in the xylem) or of carbon compounds (as in the phloem), but it is likely that they may have evolved as modifications of xylem and phloem initials in early land plants.

Tissue Types

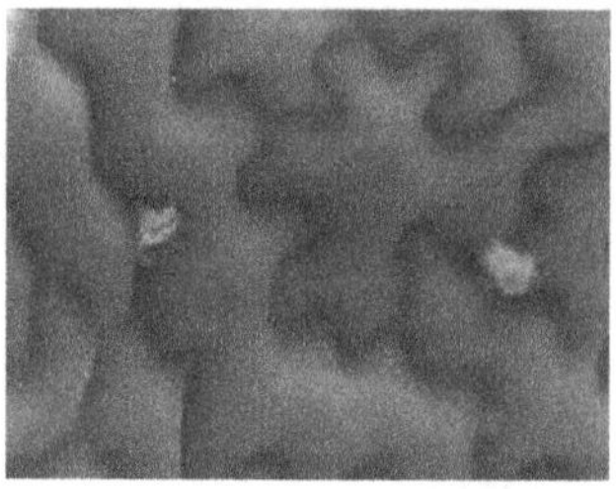

cells of Arabidopsis thalianaepidermis

The major classes of cells differentiate from undifferentiated meristematic cells (analogous to the stem cells of animals) to form the tissue structures of roots, stems, leaves, flowers, and reproductive structures.

Xylem cells are elongated cells with lignified secondary thickening of the cell walls. Xylem cells are specialised for conduction of water, and first appeared in plants during their transition to land in the Silurian period more than 425 million years ago (see Cooksonia). The possession of xylem defines the vascular plants or Tracheophytes. Xylem tracheids are pointed, elongated xylem cells, the simplest of which have continuous primary cell walls and lignified secondary wall thickenings in the form of rings, hoops, or reticulate networks. More complex tracheids with valve-like perforations called bordered pits characterise the gymnosperms. The ferns and other pteridophytes and the gymnosperms have only xylem tracheids, while the angiosperms also have xylem vessels. Vessel members are hollow xylem cells without end walls that are aligned end-to-end so as to form long continuous tubes. The bryophytes lack true xylem cells, but their sporophytes have a water-conducting tissue known as the hydrome that is composed of elongated cells of simpler construction.

Phloem is a specialised tissue for food transport in higher plants. Phloem cells mainly transport sucrose along pressure gradients generated by osmosis. This phenomenon is called translocation. Phloem consists of two cell types, the sieve tubes and the intimately associated companion cells. The sieve tube elements lack nuclei and ribosomes, and their metabolism and functions are regulated by the adjacent nucleate companion cells. Sieve tubes are joined end-to-end with perforate end-plates between known as sieve plates, which allow transport of photosynthate between the sieve elements. The companion cells, connected to the sieve tubes via plasmodesmata, are responsible for loading the phloem with sugars. The bryophytes lack phloem, but moss sporophytes have a simpler tissue with analogous function known as the leptome.

This is an electron micrograph of the epidermal cells of a Brassica chinensis leaf. The stomates are also visible.

Plant epidermal cells are specialised parenchyma cells covering the external surfaces of leaves, stems and roots. The epidermal cells of aerial organs arise from the superficial layer of cells known as the tunica (L1 and L2 layers) that covers the plant shoot apex, whereas the cortex and vascular tissues arise from innermost layer of the shoot apex known as the corpus (L3 layer). The epidermis of roots originates from the layer of cells immediately beneath the root cap.

The epidermis of all aerial organs, but not roots, is covered with a cuticle made of the polyester cutin and/or the hydrocarbon polymer cutan with a superficial layer of epicuticular waxes. The epidermal cells of the primary shoot are thought to be the only plant cells with the biochemical capacity to synthesize cutin. Several cell types may be present in the epidermis. Notable among these are the stomatal guard cells, glandular and clothing hairs or trichomes, and the root hairs of primary roots. In the shoot epidermis of most plants, only the guard cells have chloroplasts. Chloroplasts contain the green pigment chlorophyll which is needed for photosynthesis.

CELL WALL

A **cell wall** is a structural layer surrounding some types of cells, situated outside the cell membrane. It can be tough, flexible, and sometimes rigid. It provides the cell with both structural support and protection, and also acts as a filtering mechanism. Cell walls are present in most prokaryotes except mycoplasma bacteria, in algae, plants and fungi but rarely in other eukaryotes including animals. A major function is to act as pressure vessels, preventing over-expansion of the cell when water enters.

The composition of cell walls varies between species and may depend on cell type and developmental stage. The primary cell wall of land plants is composed of the polysaccharides cellulose, hemicellulose and pectin. Often, other polymers such as lignin, suberin or cutin are anchored to or embedded in plant cell walls. Algae possess walls made of glycoproteins and polysaccharides such as carrageenan and agar that are absent from land plants. In bacteria, the cell wall is composed of peptidoglycan. The cell walls of archaea have various compositions, and may be formed of glycoprotein S-layers, pseudopeptidoglycan, or polysaccharides. Fungi possess cell walls made of the glucosamine polymer chitin. Unusually, diatoms have a cell wall composed of biogenic silica.

A plant cell wall was first observed and named (simply as a "wall") by Robert Hooke in 1665. However, "the dead excrusion product of the living protoplast" was forgotten, for almost three centuries, being the subject of scientific interest mainly as a resource for industrial processing or in relation to animal or human health.

In 1804, Karl Rudolphi and J.H.F. Link proved that cells had independent cell walls. Before, it had been thought that cells shared walls and that fluid passed between them this way.

The mode of formation of the cell wall was controversial in the 19th century. Hugo von Mohl (1853, 1858) advocated the idea that the cell wall grows by apposition. Carl Nägeli (1858, 1862, 1863) believed that the growth of the wall in thickness and in area was due to a process termed intussusception. Each theory was improved in the following decades: the apposition (or lamination) theory by Eduard Strasburger (1882, 1889), and the intussusception theory by Julius Wiesner (1886).

In 1930, Ernst Münch coined the term apoplast in order to separate the "living" symplast from the "dead" plant region, the latter of which included the cell wall.

By the 1980s, some authors suggested replacing the term "cell wall", particularly as it was used for plants, with the more precise term "extracellular matrix", as used for animal cells, but others preferred the older term.

Properties

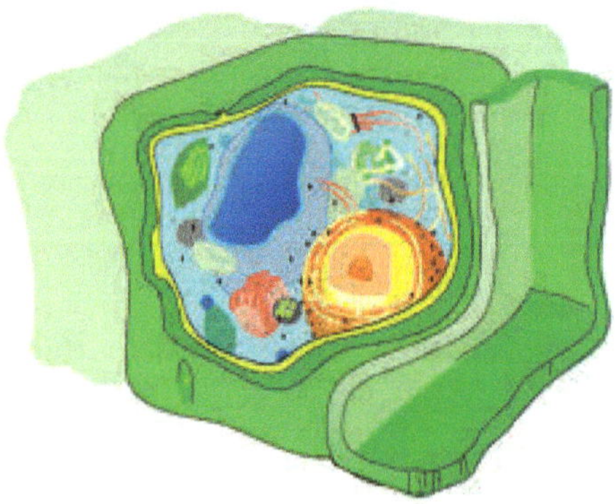

Diagram of the plant cell, with the cell wall in green.

Cell walls serve similar purposes in those organisms that possess them. They may give cells rigidity and strength, offering protection against mechanical stress. In multicellular organisms, they permit the organism to build and hold a definite shape (morphogenesis). Cell walls also limit the entry of large molecules that may be toxic to the cell. They further permit the creation of stable osmotic environments by preventing osmotic lysis and helping to retain water. Their composition, properties, and form may change during the cell cycle and depend on growth conditions.

Rigidity of Cell Walls

In most cells, the cell wall is flexible, meaning that it will bend rather than holding a fixed shape, but has considerable tensile strength. The apparent rigidity of primary plant tissues is enabled by cell walls, but is not due to the walls' stiffness. Hydraulic turgor pressure creates this rigidity, along with the wall structure. The flexibility of the cell walls is seen when plants wilt, so that the stems and leaves begin to droop, or in seaweeds that bend in water currents. As John Howland explains:

Think of the cell wall as a wicker basket in which a balloon has been inflated so that it exerts pressure from the inside. Such a basket is very rigid and resistant to mechanical damage. Thus does the prokaryote cell (and eukaryotic cell that possesses a cell wall) gain strength from a flexible plasma membrane pressing against a rigid cell wall.

The apparent rigidity of the cell wall thus results from inflation of the cell contained within. This inflation is a result of the passive uptake of water.

In plants, a **secondary cell wall** is a thicker additional layer of cellulose which increases wall rigidity. Additional layers may be formed by lignin in xylem cell walls, or suberin in cork cell walls. These compounds are rigid and waterproof, making the secondary wall stiff. Both wood and bark cells of trees have secondary walls. Other parts of plants such as the leaf stalk may acquire similar reinforcement to resist the strain of physical forces.

Permeability

The primary cell wall of most plant cells is freely permeable to small molecules including small proteins, with size exclusion estimated to be 30-60 kDa. The pH is an important factor governing the transport of molecules through cell walls.

Evolution

Cell walls evolved independently in many groups, even in the photosynthetic eukaryotes. In these lineages, the cell wall is closely related to the evolution of multicellularity, terrestrialization and vascularization.

Plant cell walls

The walls of plant cells must have sufficient tensile strength to withstand internal osmotic pressures of several times atmospheric pressure that result from the difference in solute concentration between the cell interior and external solutions.Plant cell walls vary from 0.1 to several μm in thickness.

Layers

Up to three strata or layers may be found in plant cell walls:

- The **primary cell wall**, generally a thin, flexible and extensible layer formed while the cell is growing.

- The **secondary cell wall**, a thick layer formed inside the primary cell wall after the cell is fully grown. It is not found in all cell types. Some cells, such as the conducting cells in xylem, possess a secondary wall containing lignin, which strengthens and waterproofs the wall.
- The **middle lamella**, a layer rich in pectins. This outermost layer forms the interface between adjacent plant cells and glues them together.

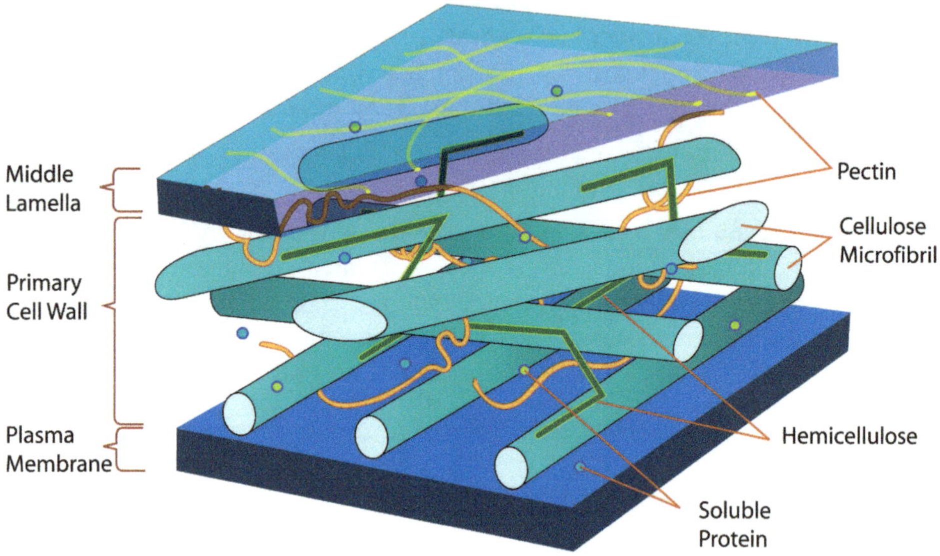

Molecular structure of the primary cell wall in plants.

Composition

In the primary (growing) plant cell wall, the major carbohydrates are cellulose, hemicellulose and pectin. The cellulose microfibrils are linked via hemicellulosic tethers to form the cellulose-hemicellulose network, which is embedded in the pectin matrix. The most common hemicellulose in the primary cell wall is xyloglucan. In grass cell walls, xyloglucan and pectin are reduced in abundance and partially replaced by glucuronarabinoxylan, another type of hemicellulose. Primary cell walls characteristically extend (grow) by a mechanism called acid growth, which involves turgor-driven movement of the strong cellulose microfibrils within the weaker hemicellulose/pectin matrix, catalyzed by expansin proteins. The outer part of the primary cell wall of the plant epidermis is usually impregnated with cutin and wax, forming a permeability barrier known as the plant cuticle.

Secondary cell walls contain a wide range of additional compounds that modify their mechanical properties and permeability. The major polymers that make up wood (largely secondary cell walls) include:

- cellulose, 35-50%

- xylan, 20-35%, a type of hemicellulose
- lignin, 10-25%, a complex phenolic polymer that penetrates the spaces in the cell wall between cellulose, hemicellulose and pectin components, driving out water and strengthening the wall.

Additionally, structural proteins (1-5%) are found in most plant cell walls; they are classified as hydroxyproline-rich glycoproteins (HRGP), arabinogalactan proteins (AGP), glycine-rich proteins (GRPs), and proline-rich proteins (PRPs). Each class of glycoprotein is defined by a characteristic, highly repetitive protein sequence. Most are glycosylated, contain hydroxyproline (Hyp) and become cross-linked in the cell wall. These proteins are often concentrated in specialized cells and in cell corners. Cell walls of the epidermis may contain cutin. The Casparian strip in the endodermis roots and cork cells of plant bark contain suberin. Both cutin and suberin are polyesters that function as permeability barriers to the movement of water. The relative composition of carbohydrates, secondary compounds and proteins varies between plants and between the cell type and age. Plant cells walls also contain numerous enzymes, such as hydrolases, esterases, peroxidases, and transglycosylases, that cut, trim and cross-link wall polymers.

Secondary walls - especially in grasses - may also contain microscopic silica crystals, which may strengthen the wall and protect it from herbivores.

Cell walls in some plant tissues also function as storage deposits for carbohydrates that can be broken down and resorbed to supply the metabolic and growth needs of the plant. For example, endosperm cell walls in the seeds of cereal grasses, nasturtium, and other species, are rich in glucans and other polysaccharides that are readily digested by enzymes during seed germination to form simple sugars that nourish the growing embryo.

Formation

The middle lamella is laid down first, formed from the cell plate during cytokinesis, and the primary cell wall is then deposited inside the middle lamella. The actual structure of the cell wall is not clearly defined and several models exist - the covalently linked cross model, the tether model, the diffuse layer model and the stratified layer model. However, the primary cell wall, can be defined as composed of cellulose microfibrils aligned at all angles. Cellulose microfibrils are produced at the plasma membrane by the cellulose synthase complex, which is proposed to be made of a hexameric rosette that contains three cellulose synthase catalytic subunits for each of the six units. Microfibrils are held together by hydrogen bonds to provide a high tensile strength. The cells are held together and share the gelatinous membrane called the middle lamella, which contains magnesium and calcium pectates (salts of pectic acid). Cells interact though plasmodesmata, which are inter-connecting channels of cytoplasm that connect to the protoplasts of adjacent cells across the cell wall.

In some plants and cell types, after a maximum size or point in development has been reached, a secondary wall is constructed between the plasma membrane and primary wall. Unlike the primary wall, the cellulose microfibrils are aligned parallel in layers, the orientation changing slightly with each additional layer so that the structure becomes helicoidal. Cells with secondary cell walls can be rigid, as in the gritty sclereid cells in pear and quince fruit. Cell to cell communication is possible through pits in the secondary cell wall that allow plasmodesmata to connect cells through the secondary cell walls.

Fungal Cell Walls

There are several groups of organisms that have been called "fungi". Some of these groups (Oomycete and Myxogastria) have been transferred out of the Kingdom Fungi, in part because of fundamental biochemical differences in the composition of the cell wall. Most true fungi have a cell wall consisting largely of chitin and other polysaccharides. True fungi do not have cellulose in their cell walls.

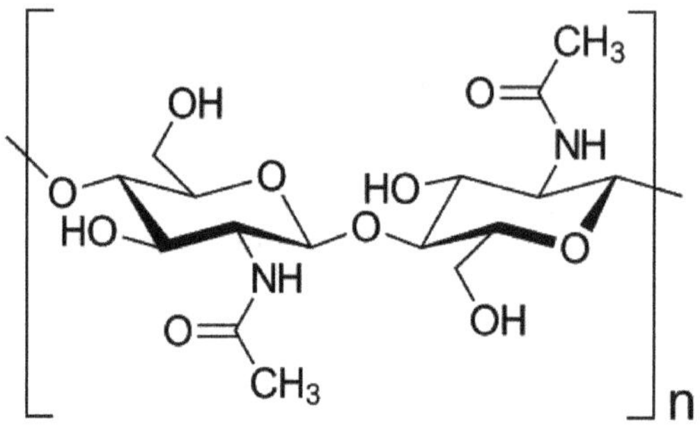

Chemical structure of a unit from a chitin polymer chain.

True Fungi

In fungi, the cell wall is the outer-most layer, external to the plasma membrane. The fungal cell wall is a matrix of three main components:

- chitin: polymers consisting mainly of unbranched chains of β-(1,4)-linked-N-Acetylglucosamine in the Ascomycota and Basidiomycota, or poly-β-(1,4)-linked-N-Acetylglucosamine (chitosan) in the Zygomycota. Both chitin and chitosan are synthesized and extruded at the plasma membrane.
- glucans: glucose polymers that function to cross-link chitin or chitosan polymers. β-glucans are glucose molecules linked via β-(1,3)- or β-(1,6)- bonds and provide rigidity to the cell wall while α-glucans are defined by α-(1,3)- and/or α-(1,4) bonds and function as part of the matrix.
- proteins: enzymes necessary for cell wall synthesis and lysis in addition to structural proteins are all present in the cell wall. Most of the structural proteins found in the cell wall are glycosylated and contain mannose, thus these proteins are called mannoproteins or mannans.

Other Eukaryotic Cell Walls

Algae

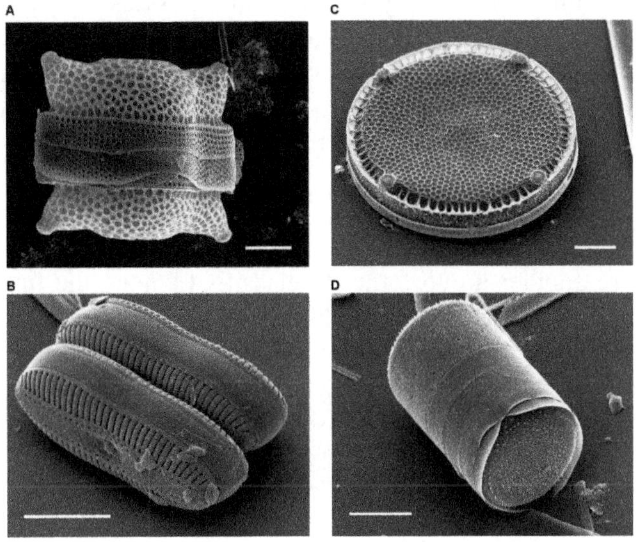

Scanning electron micrographs of diatoms showing the external appearance of the cell wall

Like plants, algae have cell walls. Algal cell walls contain either polysaccharides (such as cellulose (a glucan)) or a variety of glycoproteins (Volvocales) or both. The inclusion of additional polysaccharides in algal cells walls is used as a feature for algal taxonomy.

- Mannans: They form microfibrils in the cell walls of a number of marine green algae including those from the genera, Codium, Dasycladus, and Acetabularia as well as in the walls of some red algae, like Porphyra and Bangia.
- Xylans:
- Alginic acid: It is a common polysaccharide in the cell walls of brown algae.
- Sulfonated polysaccharides: They occur in the cell walls of most algae; those common in red algae include agarose, carrageenan, porphyran, furcelleran and funoran.

Other compounds that may accumulate in algal cell walls include sporopollenin and calcium ions.

The group of algae known as the diatoms synthesize their cell walls (also known as frustules or valves) from silicic acid (specifically orthosilicic acid, H_4SiO_4). The acid is polymerised intra-cellularly, then the wall is extruded to protect the cell. Significantly, relative to the organic cell walls produced by other groups, silica frustules require less energy to synthesize (approximately 8%), potentially a major saving on the overall cell energy budget and possibly an explanation for higher growth rates in diatoms.

In brown algae, phlorotannins may be a constituent of the cell walls.

Water molds

The group Oomycetes, also known as water molds, are saprotrophic plant pathogens like fungi. Until recently they were widely believed to be fungi, but structural and molecular evidence has led to their reclassification as heterokonts, related to autotrophic brown algae and diatoms. Unlike fungi, oomycetes typically possess cell walls of cellulose and glucans rather than chitin, although some genera (such as Achlya and Saprolegnia) do have chitin in their walls. The fraction of cellulose in the walls is no more than 4 to 20%, far less than the fraction of glucans. Oomycete cell walls also contain the amino acidhydroxyproline, which is not found in fungal cell walls.

Slime molds

The dictyostelids are another group formerly classified among the fungi. They are slime molds that feed as unicellular amoebae, but aggregate into a reproductive stalk and sporangium under certain conditions. Cells of the reproductive stalk, as well as the spores formed at the apex, possess a cellulose wall. The spore wall has been shown to possess three layers, the middle of which is composed primarily of cellulose, and the innermost is sensitive to cellulase and pronase.

Prokaryotic Cell Walls

Bacterial Cell Walls

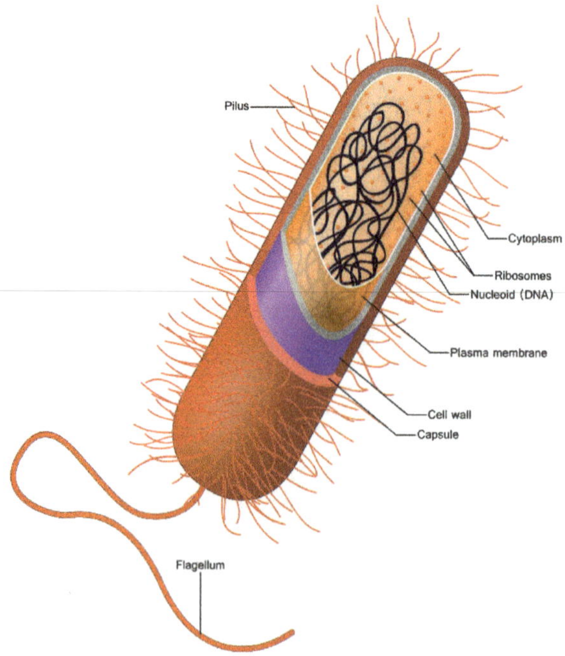

Diagram of a typical Gram-positive bacterium. The cell envelope comprises a plasma membrane, seen here in green, and a thick peptidoglycan-containing cell wall (the yellow layer). No outer lipid membrane is present, as would be the case in Gram-negative bacteria. The red layer, known as the capsule, is distinct from the cell envelope.

Further information: Cell envelope

Around the outside of the cell membrane is the bacterial cell wall. Bacterial cell walls are made of peptidoglycan (also called murein), which is made from polysaccharide chains cross-linked by unusual peptides containing D-amino acids. Bacterial cell walls are different from the cell walls of plants and fungi which are made of cellulose and chitin, respectively. The cell wall of bacteria is also distinct from that of Archaea, which do not contain peptidoglycan. The cell wall is essential to the survival of many bacteria, although L-form bacteria can be produced in the laboratory that lack a cell wall. The antibiotic penicillin is able to kill bacteria by preventing the cross-linking of peptidoglycan and this causes the cell wall to weaken and lyse. The lysozyme enzyme can also damage bacterial cell walls.

There are broadly speaking two different types of cell wall in bacteria, called Gram-positive and Gram-negative. The names originate from the reaction of cells to the Gram stain, a test long-employed for the classification of bacterial species.

Gram-positive bacteria possess a thick cell wall containing many layers of peptidoglycan and teichoic acids. In contrast, Gram-negative bacteria have a relatively thin cell wall consisting of a few layers of peptidoglycan surrounded by a second lipid membrane containing lipopolysaccharides and lipoproteins. Most bacteria have the Gram-negative cell wall and only the Firmicutes and Actinobacteria (previously known as the low G+C and high G+C Gram-positive bacteria, respectively) have the alternative Gram-positive arrangement. These differences in structure can produce differences in antibiotic susceptibility, for instance vancomycin can kill only Gram-positive bacteria and is ineffective against Gram-negative pathogens, such as Haemophilus influenzae or Pseudomonas aeruginosa.

Archaeal Cell Walls

Although not truly unique, the cell walls of Archaea are unusual. Whereas peptidoglycan is a standard component of all bacterial cell walls, all archaeal cell walls lack peptidoglycan, with the exception of one group of methanogens. In that group, the peptidoglycan is a modified form very different from the kind found in bacteria. There are four types of cell wall currently known among the Archaea.

One type of archaeal cell wall is that composed of pseudopeptidoglycan (also called pseudomurein). This type of wall is found in some methanogens, such as Methanobacterium and Methanothermus. While the overall structure of archaeal pseudopeptidoglycan superficially resembles that of bacterial peptidoglycan, there are a number of significant chemical differences. Like the peptidoglycan found in bacterial cell walls, pseudopeptidoglycan consists of polymer chains of glycan cross-linked by short peptide connections. However, unlike peptidoglycan, the sugar N-acetylmuramic acid is replaced by N-acetyltalosaminuronic acid, and the two sugars are bonded with a β,1-3 glycosidic linkage instead of β,1-4. Additionally, the cross-linking peptides are L-amino acids rather than D-amino acids as they are in bacteria.

A second type of archaeal cell wall is found in Methanosarcina and Halococcus. This type of cell wall is composed entirely of a thick layer of polysaccharides, which may be sulfated in the case of Halococcus. Structure in this type of wall is complex and as yet is not fully investigated.

A third type of wall among the Archaea consists of glycoprotein, and occurs in the hyperthermophiles, Halobacterium, and some methanogens. In Halobacterium, the proteins in the wall have a high content of acidic amino acids, giving the wall an overall negative charge. The result is an unstable structure that is stabilized by the presence of large quantities of positive sodium ions that neutralize the charge. Consequently, Halobacterium thrives only under conditions with high salinity.

In other Archaea, such as Methanomicrobium and Desulfurococcus, the wall may be composed only of surface-layer proteins, known as an S-layer. S-layers are common in bacteria, where they serve as either the sole cell-wall component or an outer layer in conjunction with polysaccharides. Most Archaea are Gram-negative, though at least one Gram-positive member is known.

Phragmoplast

The **phragmoplast** is a plant cell specific structure that forms during late cytokinesis. It serves as a scaffold for cell plate assembly and subsequent formation of a new cell wall separating the two daughter cells.

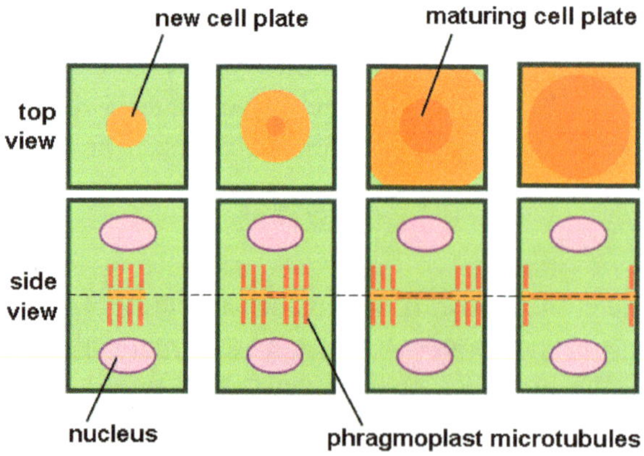

The phragmoplast is a complex assembly of microtubules (MTs), microfilaments (MFs), and endoplasmic reticulum (ER) elements, that assemble in two opposing sets perpendicular to the plane of the future cell plate during anaphase and telophase. It is initially barrel-shaped and forms from the mitotic spindle between the two daughter nuclei while nuclear envelopes reassemble around them. The cell plate initially forms as a disc between the two halves of the phragmoplast structure. While new cell plate material is added to the edges of the growing plate, the phragmoplast microtubules disappear in the center and regenerate at the edges of the growing cell plate. The two structures grow outwards until they reach the outer wall of the dividing cell. If a phragmosome was present in the cell, the phragmoplast and cell plate will grow through the space occupied by the phragmosome. They will reach the parent cell wall exactly at the position formerly occupied by the preprophase band. Phragmoplast and cell plate formation in a plant cell during cytokinesis. Left side: Phragmoplast forms and cell plate starts to assemble in the center of the cell. Towards the right: Phragmoplast enlarges in a donut-shape towards the outside of the cell, leaving behind mature cell plate in the center. The cell plate will transform into the new cell wall once cytokinesis is complete.

The microtubules and actin filaments within the phragmoplast serve to guide vesicles with cell wall material to the growing cell plate. Actin filaments are also possibly involved in guiding the phragmoplast to the site of the former preprophase band location at the parent cell wall. While the cell plate is growing, segments of smooth endoplasmic reticulum are trapped within it, later forming the plasmodesmata connecting the two daughter cells.

The phragmoplast can only be observed in Embryophytes, that is the bryophytes and vascular plants, and a few advanced green algae, specifically Coleochaete in the Division Charophyta. Some algae use another type of microtubule array, a phycoplast, during cytokinesis.

Plastid

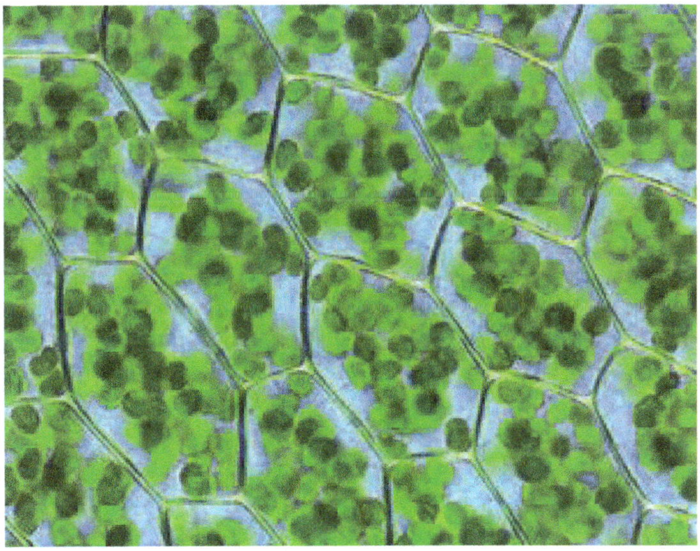

Plant cells with visible chloroplasts.

The plastid is a major double-membrane organelle found, among others, in the cells of plants and algae. Plastids are the site of manufacture and storage of important chemical compounds used by the cell. They often contain pigments used in photosynthesis, and the types of pigments present can changeor determine the cell's color. They have a common evolutionary origin and possess a double-stranded DNA molecule that is circular, like that of prokaryotic cells.

Plastids in Plants

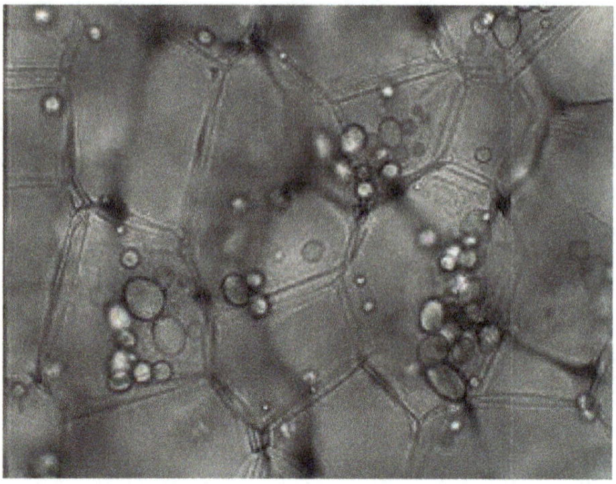

Leucoplasts in plant cells.

Those plastids that contain chlorophyll can carry out photosynthesis. Plastids can also store products like starch and can synthesise fatty acids and terpenes, which can be used for producing energy and as raw material for the synthesis of other molecules. For example, the components of the plant cuticle and its epicuticular wax are synthesized by the epidermal cells from palmitic acid, which is synthesized in the chloroplasts of the mesophyll tissue. All plastids are derived from proplastids, which are present in the meristematic regions of the plant. Proplastids and young chloroplasts commonly divide by binary fission, but more mature chloroplasts also have this capacity.

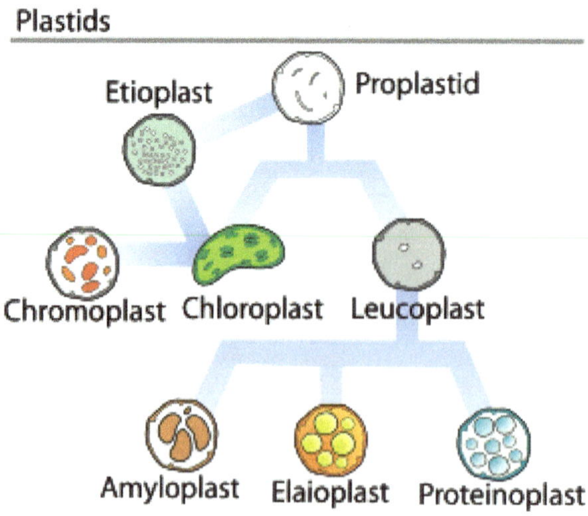

In plants, plastids may differentiate into several forms, depending upon which function they play in the cell. Undifferentiated plastids (proplastids) may develop into any of the following variants:

- Chloroplasts green plastids: for photosynthesis; see also etioplasts, the predecessors of chloroplasts
- Chromoplasts coloured plastids: for pigment synthesis and storage
- Gerontoplasts: control the dismantling of the photosynthetic apparatus during plant senescence
- Leucoplasts colourless plastids: for monoterpene synthesis; leucoplasts sometimes differentiate into more specialized plastids:
    - Amyloplasts: for starch storage and detecting gravity
    - Elaioplasts: for storing fat
    - Proteinoplasts: for storing and modifying protein
    - Tannosomes: for synthesizing and producing tannins and polyphenols

Depending on their morphology and function, plastids have the ability to differentiate, or redifferentiate, between these and other forms.

Each plastid creates multiple copies of a circular 75–250 kilobase plastome. The number of genome copies per plastid is variable, ranging from more than 1000 in rapidly dividing cells, which, in general, contain few plastids, to 100 or fewer in mature cells, where plastid divisions have given rise to a large number of plastids. The plastome contains about 100 genes encoding ribosomal and transfer ribonucleic acids (rRNAs and tRNAs) as well as proteins involved in photosynthesis and plastid gene transcription and translation. However, these proteins only represent a small fraction of the total protein set-up necessary to build and maintain the structure and function of a particular type of plastid. Plant nuclear genes encode the vast majority of plastid proteins, and the expression of plastid genes and nuclear genes is tightly co-regulated to coordinate proper development of plastids in relation to cell differentiation.

Plastid DNA exists as large protein-DNA complexes associated with the inner envelope membrane and called 'plastid nucleoids'. Each nucleoid particle may contain more than 10 copies of the plastid DNA. The proplastid contains a single nucleoid located in the centre of the plastid. The developing plastid has many nucleoids, localized at the periphery of the plastid, bound to the inner envelope membrane. During the development of proplastids to chloroplasts, and when plastids convert from one type to another, nucleoids change in morphology, size and location within the organelle. The remodelling of nucleoids is believed to occur by modifications to the composition and abundance of nucleoid proteins.

Many plastids, particularly those responsible for photosynthesis, possess numerous internal membrane layers.

In plant cells, long thin protuberances called stromules sometimes form and extend from the main plastid body into the cytosol and interconnect several plastids. Proteins, and presumably smaller molecules, can move within stromules. Most cultured cells that are relatively large compared to other plant cells have very long and abundant stromules that extend to the cell periphery. In 2014, evidence of possible plastid genome loss was found in Rafflesia lagascae, a non-photosynthetic parasitic flowering plant, and in Polytomella, a genus of non-photosynthetic green algae. Extensive searches for plastid genes in both

Rafflesia and Polytomella yielded no results, however the conclusion that their plastomes are entirely missing is still controversial.Some scientists argue that plastid genome loss is unlikely since even non-photosynthetic plastids contain genes necessary to complete various biosynthetic pathways, such as heme biosynthesis.

Plastids in Algae

In algae, the term leucoplast is used for all unpigmented plastids and their function differs from the leucoplasts of plants. Etioplasts, amyloplasts and chromoplasts are plant-specific and do not occur in algae. Plastids in algae and hornworts may also differ from plant plastids in that they contain pyrenoids.

Glaucocystophytic algae contain muroplasts, which are similar to chloroplasts except that they have a peptidoglycan cell wall that is similar to that of prokaryotes. Rhodophytic algae contain rhodoplasts, which are red chloroplasts that allow the algae to photosynthesise to a depth of up to 268 m. The chloroplasts of plants differ from the rhodoplasts of red algae in their ability to synthesize starch, which is stored in the form of granules within the plastids. In red algae, floridean starch is synthesized and stored outside the plastids in the cytosol.

Inheritance of Plastids

Most plants inherit the plastids from only one parent. In general, angiosperms inherit plastids from the female gamete, whereas many gymnosperms inherit plastids from the male pollen. Algae also inherit plastids from only one parent. The plastid DNA of the other parent is, thus, completely lost.

In normal intraspecific crossings (resulting in normal hybrids of one species), the inheritance of plastid DNA appears to be quite strictly 100% uniparental. In interspecific hybridisations, however, the inheritance of plastids appears to be more erratic. Although plastids inherit mainly maternally in interspecific hybridisations, there are many reports of hybrids of flowering plants that contain plastids of the father. Approximately 20% of angiosperms, including alfalfa (Medicago sativa), normally show biparental inheritance of plastids.

Origin of Plastids

Plastids are thought to have originated from endosymbiotic cyanobacteria. This symbiosis evolved around 1.5 billion years ago and enabled eukaryotes to carry out oxygenic photosynthesis. Three evolutionary lineages have since emerged in which the plastids are named differently: chloroplasts in green algae and plants, rhodoplasts in red algae and muroplasts in the glaucophytes. The plastids differ both in their pigmentation and in their ultrastructure. For example, chloroplasts have lost all phycobilisomes, the light harvesting complexes found in cyanobacteria, red algae and glaucophytes, but instead contain stroma and grana thylakoids, structures found only in plants and closely related green algae. The glaucocystophycean plastid — in contrast to chloroplasts and rhodoplasts — is still surrounded by the remains of the cyanobacterial cell wall. All these primary plastids are surrounded by two membranes.

Complex plastids start by secondary endosymbiosis (where a eukaryotic organism engulfs another eukaryotic organism that contains a primary plastid resulting in its endosymbiotic fixation), when a eukaryote engulfs a red or green alga and retains the algal plastid, which is typically surrounded by more than two membranes. In some cases these plastids may be reduced in their metabolic and/or photosynthetic capacity. Algae with complex plastids derived by secondary endosymbiosis of a red alga include the heterokonts, haptophytes, cryptomonads, and most dinoflagellates (= rhodoplasts). Those that endosymbiosed a green alga include the euglenids and chlorarachniophytes (= chloroplasts). The Apicomplexa, a phylum of obligate parasitic protozoa including the causative agents of malaria (Plasmodium spp.), toxoplasmosis (Toxoplasma gondii), and many other human or animal diseases also harbor a complex plastid (although this organelle has been lost in some apicomplexans, such as Cryptosporidium parvum, which causes cryptosporidiosis). The 'apicoplast' is no longer capable of photosynthesis, but is an essential organelle, and a promising target for antiparasitic drug development.

Some dinoflagellates and sea slugs, in particular of the genus Elysia, take up algae as food and keep the plastid of the digested alga to profit from the photosynthesis; after a while, the plastids are also digested. This process is known as kleptoplasty, from the Greek, kleptes, thief.

Plasmodesma

Plasmodesmata (singular: **plasmodesma**) are microscopic channels which traverse the cell walls of plant cells and some algal cells, enabling transport and communication between them. Plasmodesmata evolved independently in several lineages,and species that have these structures include members of the Charophyceae, Charales, Coleochaetales and Phaeophyceae (which are all algae), as well as all embryophytes, better known as land plants. Unlike animal cells, almost every plant cell is surrounded by a polysaccharide cell wall.

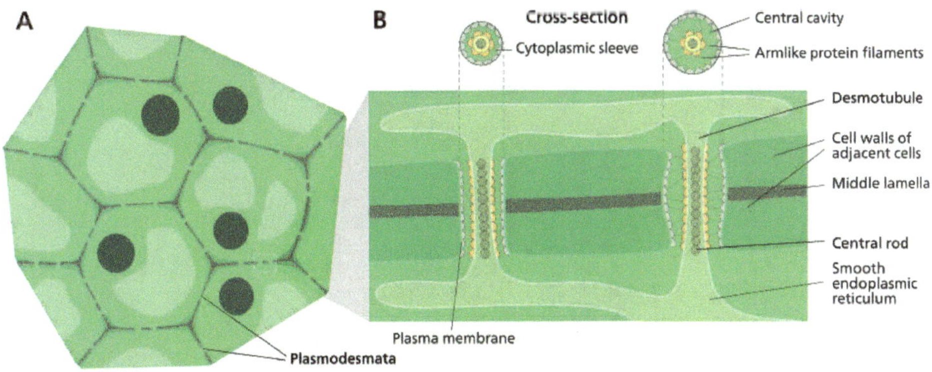

Diagram of some plasmodesmata

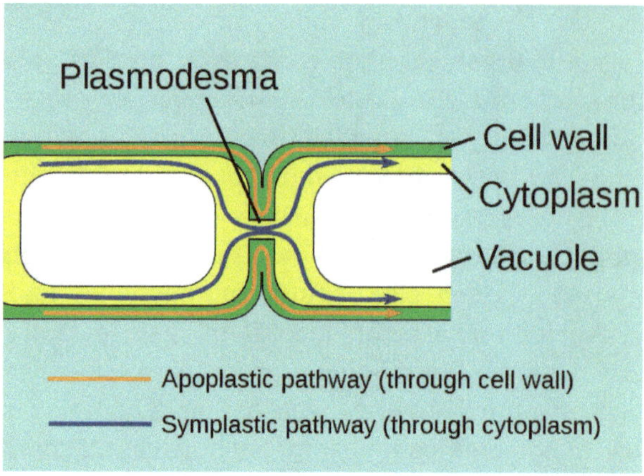

Plasmodesmata allow molecules to travel between plant cells through the symplastic **pathway**

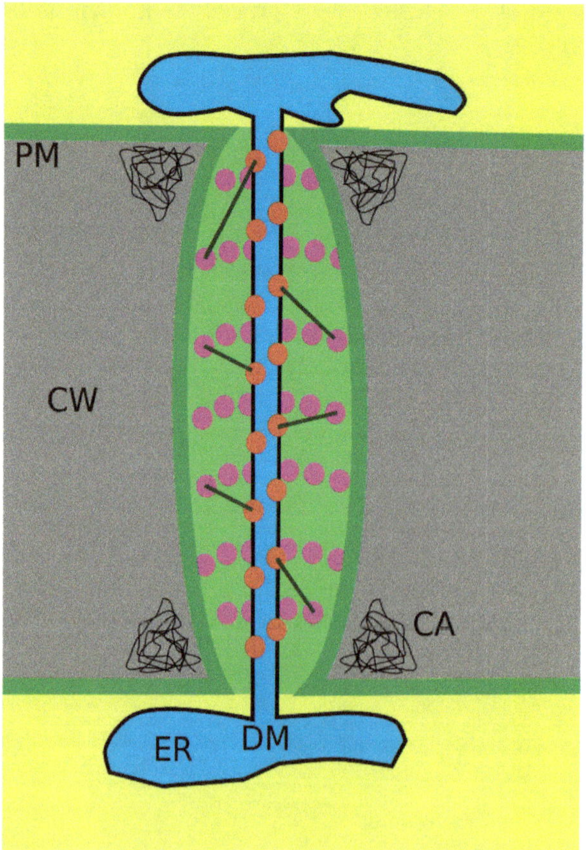

The structure of a primary plasmodesma. CW=Cell wall **CA**=Callose **PM**=Plasma membrane **ER**=Endoplasmic reticulum **DM=Desmotubule Red circles**=Actin **Purple circles and spokes**=Other unidentified proteins

Neighbouring plant cells are therefore separated by a pair of cell walls and the intervening middle lamella, forming an extracellular domain known as the apoplast. Although cell walls are permeable to small soluble proteins and other solutes, plasmodesmata enable direct, regulated, symplastic intercellular transport of substances between cells. There are two forms of plasmodesmata: primary plasmodesmata, which are formed during cell division, and secondary plasmodesmata, which can form between mature cells.

Similar structures, called gap junctions and membrane nanotubes, interconnect animal cells and stromules form between plastids in plant cells.

Formation

Primary plasmodesmata are formed when portions of the endoplasmic reticulum are trapped across the middle lamella as new cell wall is laid down between two newly divided plant cells and these eventually become the cytoplasmic connections between cells. Here the wall is not thickened further, and depressions or thin areas known as pits are formed in the walls. Pits normally pair up between adjacent cells. Plasmodesmata can also be inserted into existing cell walls between non-dividing cells (secondary plasmodesmata)

Structure

Plasmodesmatal Plasma Membrane

A typical plant cell may havc bctween 10^3 and 10^5 plasmodesmata connecting it with adjacent cells equating to between 1 and 10 per μm^2. Plasmodesmata are approximately 50-60 nm in diameter at the midpoint and are constructed of three main layers, the plasma membrane, the cytoplasmic sleeve, and the desmotubule. They can transverse cell walls that are up to 90 nm thick.

The plasma membrane portion of the plasmodesma is a continuous extension of the cell membrane or plasmalemma and has a similar phospholipid bilayer structure.

Cytoplasmic Sleeve

The cytoplasmic sleeve is a fluid-filled space enclosed by the plasmalemma and is a continuous extension of the cytosol. Trafficking of molecules and ions through plasmodesmata occurs through this space. Smaller molecules (e.g. sugars and amino acids) and ions can easily pass through plasmodesmata by diffusion without the need for additional chemical energy. Larger molecules, including proteins (for example Green fluorescent protein) and RNA, can also pass through the cytoplasmic sleeve diffusively. Plasmodesmatal transport of some larger molecules is facilitated by mechanisms that are currently unknown. One mechanism of regulation of the permeability of plasmodesmata is the accumulation of the polysaccharidecallose around the neck region to form a collar, thereby reducing the diameter of the pore available for transport of substances.

Desmotubule

The desmotubule is a tube of appressed (flattened) endoplasmic reticulum that runs between two adjacent cells. Some molecules are known to be transported through this channel, but it is not thought to be the main route for plasmodesmatal transport.

Around the desmotubule and the plasma membrane areas of an electron dense material have been seen, often joined together by spoke-like structures that seem to split the plasmodesma into smaller channels. These structures may be composed of myosin and actin, which are part of the cell's cytoskeleton. If this is the case these proteins could be used in the selective transport of large molecules between the two cells.

Transport

Plasmodesmata have been shown to transport proteins (including transcription factors), short interfering RNA, messenger RNA, viroids, and viral genomes from cell to cell. One example of a viral movement proteins is the tobacco mosaic virus MP-30. MP-30 is thought to bind to the virus's own genome and shuttle it from infected cells to uninfected cells through plasmodesmata. Flowering Locus T protein moves from leaves to the shoot apical meristem through plasmodesmata to initiate flowering.

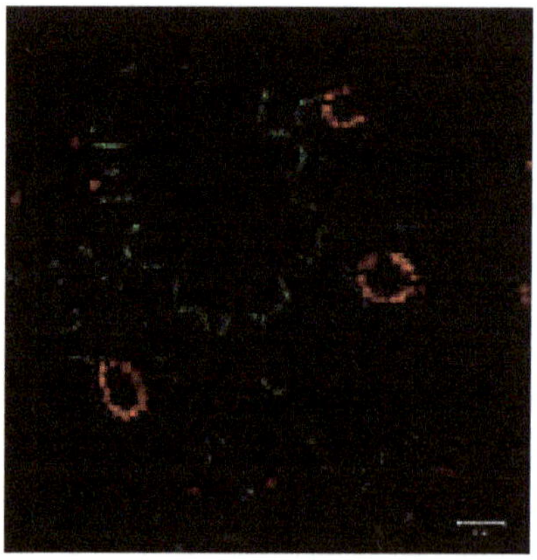

Tobacco mosaic virus movement protein 30 localizes to plasmodesmata.

Plasmodesmata are also used by cells in phloem, and symplastic transport is used to regulate the sieve-tube cells by the companion cells.

The size of molecules that can pass through plasmodesmata is determined by the size exclusion limit. This limit is highly variable and is subject to active modification. For example, MP-30 is able to increase the size exclusion limit from 700 Daltons to 9400 Daltons thereby aiding its movement through a plant. Also, increasing calcium concentrations in the cytoplasm, either by injection or by cold-induction, has been shown to constrict the opening of surrounding plasmodesmata and limit transport.

Several models for possible active transport through plasmodesmata exist. It has been suggested that such transport is mediated by interactions with proteins localized on the desmotubule, and/or by chaperones partially unfolding proteins, allowing them to fit through the narrow passage. A similar mechanism may be involved in transporting viral nucleic acids through the plasmodesmata.

Vacuole

A **vacuole** is a membrane-bound organelle which is present in all plant and fungal cells and some protist, animal and bacterial cells.Vacuoles are essentially enclosed compartments which are filled with water containing inorganic and organic molecules including enzymes in solution, though in certain cases they may contain solids which have been engulfed. Vacuoles are formed by the fusion of multiple membrane vesicles and are effectively just larger forms of these. The organelle has no basic shape or size; its structure varies according to the needs of the cell.

The function and significance of vacuoles varies greatly according to the type of cell in which they are present, having much greater prominence in the cells of plants, fungi and certain protists than those of animals and bacteria. In general, the functions of the vacuole include:

- Isolating materials that might be harmful or a threat to the cell
- Containing waste products
- Containing water in plant cells
- Maintaining internal hydrostatic pressure or turgor within the cell
- Maintaining an acidic internal pH
- Containing small molecules
- Exporting unwanted substances from the cell
- Allows plants to support structures such as leaves and flowers due to the pressure of the central vacuole
- By increasing in size, allows the germinating plant or its organs (such as leaves) to grow very quickly and using up mostly just water.
- In seeds, stored proteins needed for germination are kept in ‹protein bodies', which are modified vacuoles.

Vacuoles also play a major role in autophagy, maintaining a balance between biogenesis (production) and degradation (or turnover), of many substances and cell structures in certain organisms. They also aid in the lysis and recycling of misfolded proteins that have begun to build up within the cell. Thomas Boller and others proposed that the vacuole participates in the destruction of invading bacteria and Robert B Mellor proposed organ-specific forms have a role in 'housing' symbiotic bacteria. In protists, vacuoles have the additional function of storing food which has been absorbed by the organism and assisting in the digestive and waste management process for the cell.

The vacuole probably evolved several times independently, even within the Viridiplantae.

Discovery

Contractile vacuoles ("stars") were first observed by Spallanzani (1776) in protozoa, although mistaken for respiratory organs. Dujardin (1841) named these "stars" as vacuoles. In 1842, Schleiden applied the term for plant cells, to distinguish the structure with cell sap from the rest of the protoplasm.

In 1885, de Vries named the vacuoule membrane as tonoplast.

Bacteria

Large vacuoles are found in three genera of filamentous sulfur bacteria, the Thioploca, Beggiatoa and Thiomargarita. The cytosol is extremely reduced in these genera and the vacuole can occupy between 40–98% of the cell. The vacuole contains high concentrations of nitrate ions and is therefore thought to be a storage organelle.

Gas vacuoles, which are freely permeable to gas, are present in some species of Cyanobacteria. They allow the bacteria to control their buoyancy.

Plants

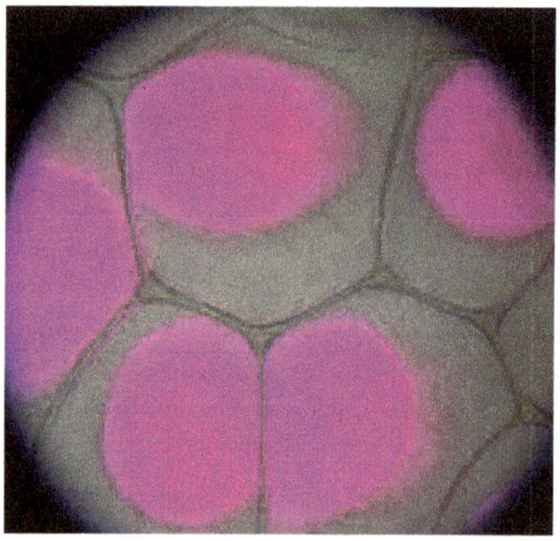

The anthocyanin-storing vacuoles of Rhoeo spathacea, a spiderwort, in cells that have plasmolyzed

Most mature plant cells have one large vacuole that typically occupies more than 30% of the cell's volume, and that can occupy as much as 80% of the volume for certain cell types and conditions. Strands of cytoplasm often run through the vacuole.

A vacuole is surrounded by a membrane called the **tonoplast** (word origin: Gk tón(os) + -o-, meaning "stretching", "tension", "tone" + comb. form repr. Gk plastós formed, molded) and filled with **cell sap**. Also called the **vacuolar membrane**, the tonoplast is the cytoplasmic membrane surrounding a vacuole, separating the vacuolar contents from the cell's cytoplasm. As a membrane, it is mainly involved in regulating the movements of ions around the cell, and isolating materials that might be harmful or a threat to the cell.

Transport of protons from the cytosol to the vacuole stabilizes cytoplasmic pH, while making the vacuolar interior more acidic creating a proton motive force which the cell can use to transport nutrients into or out of the vacuole. The low pH of the vacuole also allows degradative enzymes to act. Although single large vacuoles are most common, the size and number of vacuoles may vary in different tissues and stages of development. For example, developing cells in the meristems contain small provacuoles and cells of the vascular cambium have many small vacuoles in the winter and one large one in the summer.

Aside from storage, the main role of the central vacuole is to maintain turgor pressure against the cell wall. Proteins found in the tonoplast (aquaporins) control the flow of water into and out of the vacuole through active transport, pumping potassium (K^+) ions into and out of the vacuolar interior. Due to osmosis, water will diffuse into the vacuole, placing pressure on the cell wall. If water loss leads to a significant decline in turgor pressure, the cell will plasmolyze. Turgor pressure exerted by vacuoles is also required for cellular elongation: as the cell wall is partially degraded by the action of expansins, the less rigid wall is expanded by the pressure coming from within the vacuole. Turgor pressure exerted by the vacuole is also essential in supporting plants in an upright position. Another function of a central vacuole is that it pushes all contents of the cell's cytoplasm against the cellular membrane, and thus keeps the chloroplasts closer to light. Most plants store chemicals in the vacuole that react with chemicals in the cytosol. If the cell is broken, for example by a herbivore, then the two chemicals can react forming toxic chemicals. In garlic, alliin and the enzyme alliinase are normally separated but form allicin if the vacuole is broken. A similar reaction is responsible for the production of syn-propanethial-S-oxide when onions are cut.

Fungi

Vacuoles in fungal cells perform similar functions to those in plants and there can be more than one vacuole per cell. In yeast cells the vacuole is a dynamic structure that can rapidly modify its morphology. They are involved in many processes including the homeostasis of cell pH and the concentration of ions, osmoregulation, storing amino acids and polyphosphate and degradative processes. Toxic ions, such as strontium (Sr^{2+}), cobalt(II) (Co^{2+}), and lead(II) (Pb^{2+}) are transported into the vacuole to isolate them from the rest of the cell.

Animals

In animal cells, vacuoles perform mostly subordinate roles, assisting in larger processes of exocytosis and endocytosis.

Animal vacuoles are smaller than their plant counterparts but also usually greater in number. There are also animal cells that do not have any vacuoles.

Exocytosis is the extrusion process of proteins and lipids from the cell. These materials are absorbed into secretory granules within the Golgi apparatus before being transported to the cell membrane and secreted into the extracellular environment. In this capacity, vacuoles are simply storage vesicles which allow for the containment, transport and disposal of selected proteins and lipids to the extracellular environment of the cell. Endocytosis is the reverse of exocytosis and can occur in a variety of forms. Phagocytosis ("cell eating") is the process by which bacteria, dead tissue, or other bits of material visible under the microscope are engulfed by cells. The material makes contact with the cell membrane, which then invaginates. The invagination is pinched off, leaving the engulfed material in the membrane-enclosed vacuole and the cell membrane intact. Pinocytosis ("cell drinking") is essentially the same process, the difference being that the substances ingested are in solution and not visible under the microscope. Phagocytosis and pinocytosis are both undertaken in association with lysosomes which complete the breakdown of the material which has been engulfed.

Salmonella is able to survive and reproduce in the vacuoles of several mammal species after being engulfed.

TISSUES

Meristem

A **meristem** is the tissue in most plants containing undifferentiated cells (**meristematic cells**), found in zones of the plant where growth can take place.

Meristematic cells give rise to various organs of the plant and keep the plant growing. The shoot apical meristem (SAM) gives rise to organs like the leaves and flowers, while the root apical meristem (RAM) provides the meristematic cells for the future root growth. SAM and RAM cells divide rapidly and are considered indeterminate, in that they do not possess any defined end status. In that sense, the meristematic cells are frequently compared to the stem cells in animals, which have an analogous behavior and function.

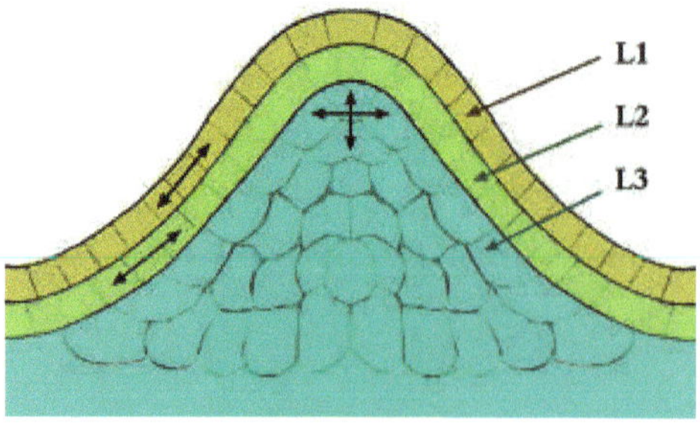

Tunica-Corpus model of the apical meristem (growing tip). The epidermal (L1) and subepidermal (L2) layers form the outer layers called the tunica. The inner L3 layer is called the corpus. Cells in the L1 and L2 layers divide in a sideways fashion, which keeps these layers distinct, whereas the L3 layer divides in a more random fashion.

The term meristem was first used in 1858 by Karl Wilhelm von Nägeli (1817–1891) in his book Beiträge zur Wissenschaftlichen Botanik ("Contributions to Scientific Botany").

It is derived from the Greek word merizein (μερίζειν), meaning to divide, in recognition of its inherent function.

In general, differentiated plant cells cannot divide or produce cells of a different type. Therefore, cell division in the meristem is required to provide new cells for expansion and differentiation of tissues and initiation of new organs, providing the basic structure of the plant body.

Meristematic cells are incompletely or not at all differentiated, and are capable of continued cellular division (youthful). Furthermore, the cells are small and protoplasm fills the cell completely. The vacuoles are extremely small. The cytoplasm does not contain differentiated plastids (chloroplasts or chromoplasts), although they are present in rudimentary form (proplastids). Meristematic cells are packed closely together without intercellular cavities. The cell wall is a very thin primary cell wall.

Maintenance of the cells requires a balance between two antagonistic processes: organ initiation and stem cell population renewal.

Apical meristems are the completely undifferentiated (indeterminate) meristems in a plant. These differentiate into three kinds of primary meristems. The primary meristems in turn produce the two secondary meristem types. These secondary meristems are also known as lateral meristems because they are involved in lateral growth.

At the meristem summit, there is a small group of slowly dividing cells, which is commonly called the central zone. Cells of this zone have a stem cell function and are essential for meristem maintenance. The proliferation and growth rates at the meristem summit usually differ considerably from those at the periphery.

Meristems also are induced in the roots of legumes such as soybean, Lotus japonicus, pea, and Medicago truncatula after infection with soil bacteria commonly called Rhizobium. Cells of the inner or outer cortex in the so-called "window of nodulation" just behind the developing root tip are induced to divide. The critical signal substance is the lipo-oligosaccharideNod-factor, decorated with side groups to allow specificity of interaction. The Nod factor receptor proteins NFR1 and NFR5 were cloned from several legumes including Lotus japonicus, Medicago truncatula and soybean (Glycine max). Regulation of nodule meristems utilizes long distance regulation commonly called "Autoregulation of Nodulation" (AON). This process involves a leaf-vascular tissue located LRR receptor kinases (LjHAR1, GmNARK and MtSUNN), CLE peptide signalling, and KAPP interaction, similar to that seen in the CLV1,2,3 system. LjKLAVIER also exhibits a nodule regulation phenotype though it is not yet known how this relates to the other AON receptor kinases.

Apical Meristems

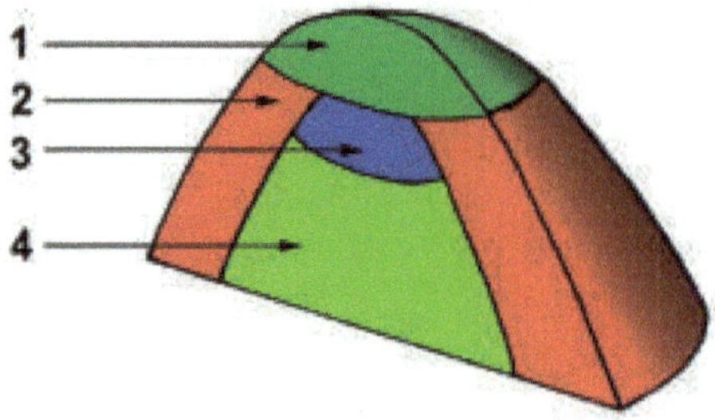

Organisation of an apical meristem (growing tip)

1 - Central zone

2 - Peripheral zone

3 - Medullary (i.e. central) meristem

4 - Medullary tissue

The number of layers varies according to plant type. In general the outermost layer is called the **tunica** while the innermost layers are the **corpus**. In monocots, the tunica determine the physical characteristics of the leaf edge and margin. In dicots, layer two of the corpus determine the characteristics of the edge of the leaf. The corpus and tunica play a critical part of the plant physical appearance as all plant cells are formed from the meristems. Apical meristems are found in two locations: the root and the stem. Some Arctic plants have an apical meristem in the lower/middle parts of the plant. It is thought that this kind of meristem evolved because it is advantageous in Arctic conditions.

Shoot Apical Meristems

The source of all above-ground organs. Cells at the shoot apical meristem summit serve as stem cells to the surrounding peripheral region, where they proliferate rapidly and are incorporated into differentiating leaf or flower primordia.

Shoot apical meristems of Crassula ovata (left). Fourteen days later, leaves have developed (right).

The shoot apical meristem is the site of most of the embryogenesis in flowering plants. Primordia of leaves, sepals, petals, stamens and ovaries are initiated here at the rate of one every time interval, called a plastochron. It is where the first indications that flower development has been evoked are manifested. One of these indications might be the loss of apical dominance and the release of otherwise dormant cells to develop as auxiliary shoot meristems, in some species in axils of primordia as close as two or three away from the apical dome. The shoot apical meristem consists of 4 distinct cell groups:

- Stem cells
- The immediate daughter cells of the stem cells
- A subjacent organising centre
- Founder cells for organ initiation in surrounding regions

The four distinct zones mentioned above are maintained by a complex signalling pathway. In Arabidopsis thaliana, 3 interacting CLAVATA genes are required to regulate the size of the stem cell reservoir in the shoot apical meristem by controlling the rate of cell division. CLV1 and CLV2 are predicted to form a receptor complex (of the LRR receptor-like kinase family) to which CLV3 is a ligand. CLV3 shares some homology with the ESR proteins of maize, with a short 14 amino acid region being conserved between the proteins. Proteins that contain these conserved regions have been grouped into the CLE family of proteins.

CLV1 has been shown to interact with several cytoplasmic proteins that are most likely involved in downstream signalling. For example, the CLV complex has been found to be associated with Rho/Rac small GTPase-related proteins. These proteins may act as an intermediate between the CLV complex and a mitogen-activated protein kinase (MAPK), which is often involved in signalling cascades. KAPP is a kinase-associated protein phosphatase that has been shown to interact with CLV1.KAPP is thought to act as a negative regulator of CLV1 by dephosphorylating it.

Another important gene in plant meristem maintenance is WUSCHEL (shortened to WUS), which is a target of CLV signalling. WUS is expressed in the cells below the stem cells of the meristem and its presence prevents the differentiation of the stem cells. CLV1 acts to promote cellular differentiation by repressing WUS activity outside of the central zone containing the stem cells. SHOOT MERISTEMLESS (STM) also acts to prevent the differentiation of stem cells by repressing the expression of MYB genes that are involved in cellular differentiation.

Root Apical Meristem

Unlike the shoot apical meristem, the root apical meristem produces cells in two dimensions. It harbors two pools of stem cells around an organizing center called the quiescent center (QC) cells and together produce most of the cells in an adult root. At its apex, the root meristem is covered by the root cap, which protects and guides its growth trajectory. Cells are continuously sloughed off the outer surface of the root cap. The QC cells are characterized by their low mitotic activity. Evidence suggests that the QC maintains the surrounding stem cells by preventing their differentiation, via signal(s) that are yet to be discovered. This allows a constant supply of new cells in the meristem required for continuous root growth. Recent findings indicate that QC can also act as a reservoir of stem cells to replenish whatever is lost or damaged. Root apical meristem and tissue patterns become established in the embryo in the case of the primary root, and in the new lateral root primordium in the case of secondary roots.

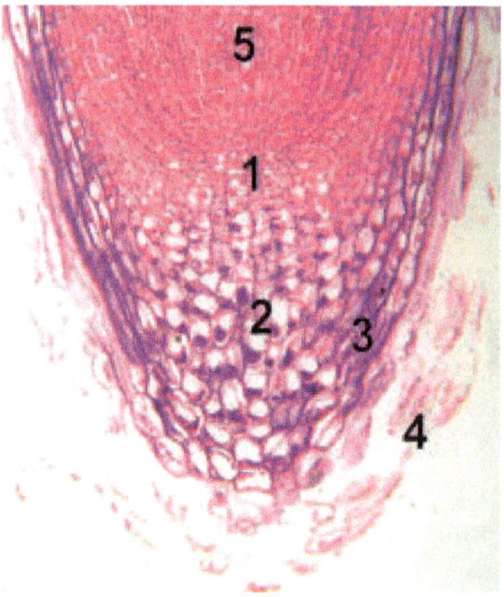

10x microscope image of root tip with meristem

1 - quiescent center

2 - calyptrogen (live rootcap cells)

3 – rootcap

4 - sloughed off dead rootcap cells

5 - procambium

Intercalary Meristem

In angiosperms, intercalary meristems occur only in monocot (in particular, grass) stems at the base of nodes and leaf blades. Horsetails also exhibit intercalary growth. Intercalary meristems are capable of cell division, and they allow for rapid growth and regrowth of many monocots. Intercalary meristems at the nodes of bamboo allow for rapid stem elongation, while those at the base of most grass leaf blades allow damaged leaves to rapidly regrow. This leaf regrowth in grasses evolved in response to damage by grazing herbivores, but is more familiar to us in response to lawnmowers.

Floral Meristem

When plants begin the developmental process known as flowering, the shoot apical meristem is transformed into an inflorescence meristem, which goes on to produce the floral meristem, which produces the sepals, petals, stamens, and carpels of the flower. In contrast to vegetative apical meristems and some exflorescence meristems, floral meristems cannot continue to grow indefinitely. Their future growth is limited to the flower with a particular size and form. The transition from shoot meristem to floral meristem requires floral meristem identity genes, that both specify the floral organs and cause the termination of the production of stem cells. AGAMOUS (AG) is a floral homeotic gene required for floral meristem termination and necessary for proper development of the stamens and carpels. AG is necessary to prevent the conversion of floral meristems to inflorescence shoot meristems, but is not involved in the transition from shoot to floral meristem. AG is turned on by the floral meristem identity gene LEAFY (LFY) and WUS and is restricted to the centre of the floral meristem or the inner two whorls. This way floral identity and region specificity is achieved. WUS activates AG by binding to a consensus sequence in the AG's second intron and LFY binds to adjacent recognition sites. Once AG is activated it represses expression of WUS leading to the termination of the meristem.

Through the years, scientists have manipulated floral meristems for economic reasons. An example is the mutant tobacco plant "Maryland Mammoth." In 1936, the department of agriculture of Switzerland performed several scientific tests with this plant. "Maryland Mammoth" is peculiar in that it grows much faster than other tobacco plants.

Apical Dominance

Apical dominance is phenomenon where one meristem prevents or inhibits the growth of other meristems. As a result, the plant will have one clearly defined main trunk. For example, in trees, the tip of the main trunk bears the dominant meristem. Therefore, the tip of the trunk grows rapidly and is not shadowed by branches. If the dominant meristem is cut off, one or more branch tips will assume dominance. The branch will start growing faster and the new growth will be vertical. Over the years, the branch may begin to look more and more like an extension of the main trunk. Often several branches will exhibit this behaviour after the removal of apical meristem, leading to a bushy growth.

The mechanism of apical dominance is based on the plant hormone auxin. It is produced in the apical meristem and transported towards the roots in the cambium. If apical dominance is complete, it prevents any branches from forming as long as the apical meristem is active. If the dominance is incomplete, side branches will develop.

Recent investigations into apical dominance and the control of branching have revealed a new plant hormone family termed strigolactones. These compounds were previously known to be involved in seed germination and communication with mycorrhizal fungi and are now shown to be involved in inhibition of branching.

Diversity in Meristem Architectures

Is the mechanism of being indeterminate conserved in the SAMs of the plant world? The SAM contains a population of stem cells that also produce the lateral meristems while the stem elongates. It turns out that the mechanism of regulation of the stem cell number might indeed be evolutionarily conserved. The CLAVATA gene CLV2 responsible for maintaining the stem cell population in Arabidopsis thaliana is very closely related to the Maize gene FASCIATED EAR 2(FEA2) also involved in the same function. Similarly, in Rice, the FON1-FON2 system seems to bear a close relationship with the CLV signaling system in Arabidopsis thaliana. These studies suggest that the regulation of stem cell number, identity and differentiation might be an evolutionarily conserved mechanism in monocots, if not in angiosperms. Rice also contains another genetic system distinct from FON1-FON2, that is involved in regulating stem cell number. This example underlines the innovation that goes about in the living world all the time.

Role of the KNOX-family genes

Genetic screens have identified genes belonging to the KNOX family in this function. These genes essentially maintain the stem cells in an undifferentiated state. The KNOX family has undergone quite a bit of evolutionary diversification, while keeping the overall mechanism more or less similar. Members of the KNOX family have been found in plants as diverse as Arabidopsis thaliana, rice, barley and tomato. KNOX-like genes are also present in some algae, mosses, ferns and gymnosperms. Misexpression of these genes leads to formation of interesting morphological features. For example, among members of Antirrhinae, only the species of genus Antirrhinum lack a structure called spur in the floral region. A spur is considered an evolutionary innovation because it defines pollinator specificity and attraction. Researchers carried out transposon mutagenesis in Antirrhinum majus, and saw that some insertions led to formation of spurs that were very similar to the other members of Antirrhinae, indicating that the loss of spur in wild Antirrhinum majus populations could probably be an evolutionary innovation.

Note the long spur of the above flower. Spurs attract pollinators and confer pollinator specificity. (Flower:Linaria dalmatica)

Complex leaves of C. hirsuta are a result of KNOX gene expression

The KNOX family has also been implicated in leaf shape evolution (See below for a more detailed discussion). One study looked at the pattern of KNOX gene expression in A. thaliana, that has simple leaves and Cardamine hirsuta, a plant having complex leaves. In A. thaliana, the KNOX genes are completely turned off in leaves, but in C.hirsuta, the expression continued, generating complex leaves. Also, it has been proposed that the mechanism of KNOX gene action is conserved across all vascular plants, because there is a tight correlation between KNOX expression and a complex leaf morphology.

Primary Meristems

Apical meristems may differentiate into three kinds of primary meristem:

- **Protoderm**: lies around the outside of the stem and develops into the epidermis.
- **Procambium**: lies just inside of the protoderm and develops into primary xylem and primary phloem. It also produces the vascular cambium, and cork cambium, secondary meristems. The cork cambium further differentiates into the phelloderm (to the inside) and the phellem, or cork (to the outside). All three of these layers (cork cambium, phellem and phelloderm) constitute the periderm. In roots, the procambium can also give rise to the pericycle, which produces lateral roots in eudicots.
- **Ground meristem**: develops into the cortex and the pith. Composed of parenchyma, collenchyma and sclerenchyma cells.

These meristems are responsible for primary growth, or an increase in length or height, which were discovered by scientist Joseph D. Carr of North Carolina in 1943.

Secondary meristems

There are two types of secondary meristems, these are also called the lateral meristems because they surround the established stem of a plant and cause it to grow laterally (i.e., larger in diameter).

- Vascular cambium, which produces secondary xylem and secondary phloem. This is a process that may continue throughout the life of the plant. This is what gives rise to wood in plants. Such plants are called arborescent. This does not occur in plants that do not go through secondary growth (known as herbaceous plants).
- Cork cambium, which gives rise to the periderm, which replaces the epidermis.

Indeterminate Growth of Meristems

Though each plant grows according to a certain set of rules, each new root and shoot meristem can go on growing for as long as it is alive. In many plants, meristematic growth is potentially **indeterminate**, making the overall shape of the plant not determinate in advance. This is the **primary growth**. Primary growth leads to lengthening of the plant body and organ formation. All plant organs arise ultimately from cell divisions in the apical meristems, followed by cell expansion and differentiation. Primary growth gives rise to the apical part of many plants. The growth of nitrogen fixing nodules on legume plants such as soybean and pea is either determinate or indeterminate. Thus, soybean (or bean and Lotus japonicus) produce determinate nodules (spherical), with a branched vascular system surrounding the central infected zone. Often, Rhizobium infected cells have only small vacuoles. In contrast, nodules on pea, clovers, and Medicago truncatula are indeterminate, to maintain (at least for some time) an active meristem that yields new cells for Rhizobium infection. Thus zones of maturity exist in the nodule. Infected cells usually possess a large vacuole. The plant vascular system is branched and peripheral.

Cloning

Under appropriate conditions, each shoot meristem can develop into a complete new plant or clone. Such new plants can be grown from shoot cuttings that contain an apical meristem. Root apical meristems are not readily cloned, however. This cloning is called **asexual reproduction** or **vegetative reproduction** and is widely practiced in horticulture to mass-produce plants of a desirable genotype. This process is also known as mericloning.

Propagating through cuttings is another form of vegetative propagation that initiates root or shoot production from secondary meristematic cambial cells. This explains why basal ‹wounding' of shoot-borne cuttings often aids root formation.

Storage Organ

A **storage organ** is a part of a plant specifically modified for storage of energy (generally in the form of carbohydrates) or water. Storage organs often grow underground, where they are better protected from attack by herbivores. Plants that have an underground storage organ are called **geophytes** in the Raunkiær plant life-form classification system. Storage organs often, but not always, act as perennating organs which enable plants to survive adverse conditions (such as cold, excessive heat, lack of light or drought).

Relationship to Perennating Organ

Storage organs may act as perennating organs ('perennating' as in perennial, meaning "through the year", used in the sense of continuing beyond the year and in due course lasting for multiple years). These are used by plants to survive adverse periods in the plant's life-cycle (e.g. caused by cold, excessive heat, lack of light or drought). During these periods, parts of the plant die and then when conditions become favourable again, re-growth occurs from buds in the perennating organs. For example geophytes growing in woodland under deciduous trees (e.g. bluebells, trilliums) die back to underground storage organs during summer when tree leaf cover restricts light and water is less available.

However, perennating organs need not be storage organs. After losing their leaves, deciduous trees grow them again from ‹resting buds', which are the perennating organs of phanerophytes in the Raunkiær classification, but which do not specifically act as storage organs. Equally, storage organs need not be perennating organs. Many succulents have leaves adapted for water storage, which they retain in adverse conditions.

Underground Storage Organ

In common parlance, underground storage organs may be generically called roots, tubers, or bulbs, but to the botanist there is more specific technical nomenclature:

A harvested ginger rhizome

- True roots:
  - Tuberous root or root tuber — e.g. Dahlia
  - Storage taproot — e.g. carrot
- Modified stems:
  - Corm — e.g. Crocus
  - Stem tuber — e.g. Zantedeschia (arum lily), potato
  - Rhizome — e.g. Iris pseudacorus (yellow flag iris)
  - Pseudobulb — e.g. Pleione (windowsill orchid)
  - Caudex — e.g. Adenium (desert-rose)
- Others:
  - Storage hypocotyl (the stem of a seedling) — sometimes called a tuber, as in Cyclamen
  - Bulb (a short stem that produces fleshy scale leaves or modified leaf bases) — e.g. Lilium, Narcissus, onion

Some of the above, particularly pseudobulbs and caudices, may occur wholly or partially above ground. Intermediates and combinations of the above are also found, making classification difficult. As an example of an intermediate, the tuber of Cyclamen arises from the stem of the seedling, which forms the junction of the roots and stem of the mature plant. In some species (e.g. Cyclamen coum) roots come from the bottom of the tuber, suggesting that it is a stem tuber; in others (e.g. Cyclamen hederifolium) roots come largely from the top of the tuber, suggesting that it is a root tuber. As an example of a combination, juno irises have both bulbs and storage roots. Underground storage organs used for food may be generically called root vegetables, although this phrase should not be taken to imply that the class only includes true roots.

Other Storage Organs

Crassula arborescens, a leaf succulent

Succulents are plants which are adapted to withstand periods of drought by their ability to store moisture in specialized storage organs.

- Leaf succulents store water in their leaves, which are thus thickened, fleshy and typically covered with a waxy coating or fine hairs to reduce evaporation. They may also contain mucilagenous compounds. Some leaf succulents have leaves which are distributed along the stem in a similar fashion to non-succulent species (e.g. Crassula, Kalanchoe); their stems may also be succulent. In others, the leaves are more compact, forming a rosette (e.g. Echeveria, Aloe). Pebble-plants or living stones (e.g. Lithops, Conophytum) have reduced their leaves to just two, forming a fleshy body, only the top of which may be visible above ground.

Ferocactus pilosus (Mexican lime cactus), a stem succulent

- Stem succulents are generally either leafless or have leaves which can be quickly shed in the event of drought. Photosynthesis is then taken over by the stems. As with leaf succulents, stems may be covered with a waxy coating or fine hairs to reduce evaporation. The ribbed bodies of cacti may be an adaption to allow shrinkage and expansion with the amount of water stored. Plants of the same general form as cacti are found in other families (e.g. Euphorbia canariensis (family Euphorbiaceae), Stapelia (family Apocynaceae)).

Ground Tissue

The **ground tissue** of plants includes all tissues that are neither dermal nor vascular. It can be divided into three classes based on the nature of the cell walls. **Parenchyma** cells have thin primary walls and usually remain alive after they become mature. Parenchyma forms the "filler" tissue in the soft parts of plants. **Collenchyma** cells have thin primary walls with some areas of secondary thickening. Collenchyma provides extra structural support, particularly in regions of new growth. **Sclerenchyma** cells have thick lignified secondary walls and often die when mature. Sclerenchyma provides the main structural support to a plant.

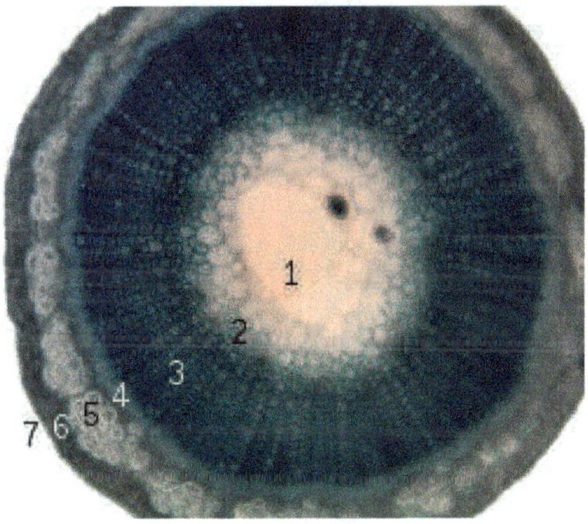

Cross-section of a flax plant stem:

- Pith
- Protoxylem
- Xylem I
- Phloem I
- Sclerenchyma (bast fibre)
- Cortex
- Epidermis

Parenchyma

Parenchyma is a versatile ground tissue that generally constitutes the "filler" tissue in soft parts of plants. It forms, among other things, the cortex and pith of stems, the cortex of roots, the mesophyll of leaves, the pulp of fruits, and the endosperm of seeds. Parenchyma cells are living cells and may remain meristematic at maturity—meaning that they are capable of cell division if stimulated. They have thin but flexible cellulose cell walls, and are generally polyhedral when close-packed, but can be roughly spherical when isolated from their neighbours. They have large central vacuoles, which allow the cells to store and regulate ions, waste products, and water. Tissue specialised for food storage is commonly formed of parenchyma cells.

Parenchyma cells have a variety of functions:

- In leaves, they form the mesophyll and are responsible for photosynthesis and the exchange of gases,parenchyma cells in the mesophyll of leaves are specialised parenchyma cells called chlorenchyma cells (parenchyma cells with chloroplasts).
- Storage of starch, protein, fats, oils and water in roots, tubers (e.g. potatoes), seed endosperm (e.g. cereals) and cotyledons (e.g. pulses and peanuts)
- Secretion (e.g. the parenchyma cells lining the inside of resin ducts)
- Wound repair and the potential for renewed meristematic activity
- Other specialised functions such as aeration (aerenchyma) provides buoyancy and helps aquatic plants in floating.
- Chlorenchyma cells carry out photosynthesis and manufacture food.

The shape of parenchyma cells varies with their function. In the spongy mesophyll of a leaf, parenchyma cells range from near-spherical and loosely arranged with large intercellular spaces, to branched or stellate, mutually interconnected with their neighbours at the ends of their arms to form a three-dimensional network, like in the red kidney bean Phaseolus vulgaris and other mesophytes. These cells, along with the epidermal guard cells of the stoma, form a system of air spaces and chambers that regulate the exchange of gases. In some works the cells of the leaf epidermis are regarded as specialised parenchymal cells, but the modern preference has long been to classify the epidermis as plant dermal tissue, and parenchyma as ground tissue. shapes of parenchyma 1= polyhedral[these cells are most poly hedral shape] 2=stellate (found in stem of plants and have well developed air spaces btween them) 3=elongated(are found in pallisade tissue of leaf) 4=lobed (are found inspongy and pallisade mesophyyll tissue of some plants)

Collenchyma

The first use of "collenchyma" was by Link (1837) who used it to describe the sticky substance on Bletia (Orchidaceae) pollen. Complaining about Link's excessive nomenclature, Schleiden (1839) stated mockingly that the term "collenchyma" could have more easily been used to describe elongated sub-epidermal cells with unevenly thickened cell walls.

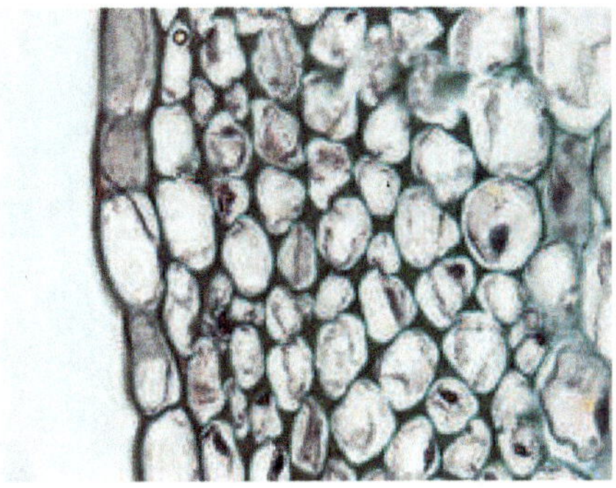

Cross section of collenchyma cells

Collenchyma tissue is composed of elongated cells with irregularly thickened walls. They provide structural support, particularly in growing shoots and leaves. Collenchyma tissue makes up things such as the resilient strands in stalks of celery. Collenchyma cells are usually living, and have only a thick primary cell wall made up of cellulose and pectin. Cell wall thickness is strongly affected by mechanical stress upon the plant. The walls of collenchyma in shaken plants (to mimic the effects of wind etc.), may be 40–100% thicker than those not shaken.

There are four main types of collenchyma:

- Angular collenchyma (thickened at intercellular contact points)
- Tangential collenchyma (cells arranged into ordered rows and thickened at the tangential face of the cell wall)
- Annular collenchyma (uniformly thickened cell walls)
- Lacunar collenchyma (collenchyma with intercellular spaces)

Collenchyma cells are most often found adjacent to outer growing tissues such as the vascular cambium and are known for increasing structural support and integrity.

Sclerenchyma

Sclerenchyma is the supporting tissue in plants. Two types of sclerenchyma cells exist: fibres and sclereids. Their cell walls consist of cellulose, hemicellulose and lignin. Sclerenchyma cells are the principal supporting cells in plant tissues that have ceased elongation. Sclerenchyma fibres are of great economic importance, since they constitute the source material for many fabrics (e.g. flax, hemp, jute, and ramie).

Unlike the collenchyma, mature sclerenchyma is composed of dead cells with extremely thick cell walls (secondary walls) that make up to 90% of the whole cell volume. The term sclerenchyma is derived from the Greek σκληρός (sklērós), meaning "hard." It is the hard, thick walls that make sclerenchyma cells important strengthening and supporting elements in plant parts that have ceased elongation. The difference between fibres and sclereids is not always clear: transitions do exist, sometimes even within the same plant.

Fibers

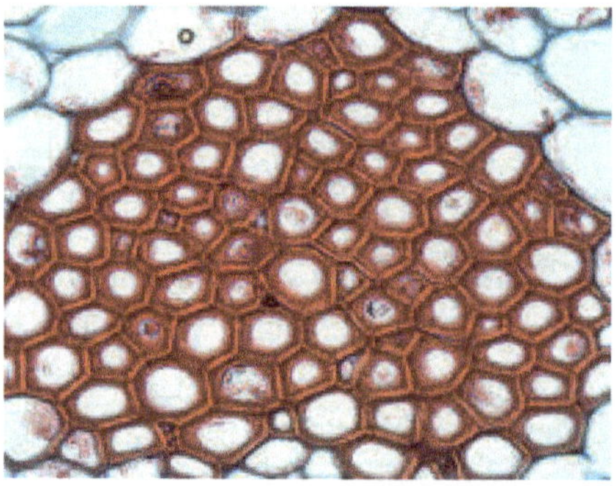

Cross section of sclerenchyma fibres

Fibers or bast are generally long, slender, so-called prosenchymatous cells, usually occurring in strands or bundles. Such bundles or the totality of a stem's bundles are colloquially called fibres. Their high load-bearing capacity and the ease with which they can be processed has since antiquity made them the source material for a number of things, like ropes, fabrics and mattresses. The fibres of flax (Linum usitatissimum) have been known in Europe and Egypt for more than 3,000 years, those of hemp (Cannabis sativa) in China for just as long. These fibres, and those of jute (Corchorus capsularis) and ramie (Boehmeria nivea, a nettle), are extremely soft and elastic and are especially well suited for the processing to textiles. Their principal cell wall material is cellulose.

Contrasting are hard fibres that are mostly found in monocots. Typical examples are the fibres of many grasses, agaves (sisal: Agave sisalana), lilies (Yucca or Phormium tenax), Musa textilis and others. Their cell walls contain, besides cellulose, a high proportion of lignin. The load-bearing capacity of Phormium tenax is as high as 20–25 kg/mm^2, the same as that of good steel wire (25 kg/mm^2), but the fibre tears as soon as too great a strain is placed upon it, while the wire distorts and does not tear before a strain of 80 kg/mm^2. The thickening of a cell wall has been studied in Linum. Starting at the centre of the fibre, the thickening layers of the secondary wall are deposited one after the other. Growth at both tips of the cell leads to simultaneous elongation. During development the layers of secondary material seem like tubes, of which the outer one is always longer and older than the next. After completion of growth, the missing parts are supplemented, so that the wall is evenly thickened up to the tips of the fibres.

Fibres usually originate from meristematic tissues. Cambium and procambium are their main centres of production. They are usually associated with the xylem and phloem of the vascular bundles. The fibres of the xylem are always lignified, while those of the phloem are cellulosic. Reliable evidence for the fibre cells' evolutionary origin from tracheids exists. During evolution the strength of the tracheid cell walls was enhanced, the ability to conduct water was lost and the size of the pits was reduced. Fibres that do not belong to the xylem are bast (outside the ring of cambium) and such fibres that are arranged in characteristic patterns at different sites of the shoot.

Sclereids

Sclereids are a reduced form of sclerenchyma cells with highly thickened, lignified walls. These have a shape of a star.

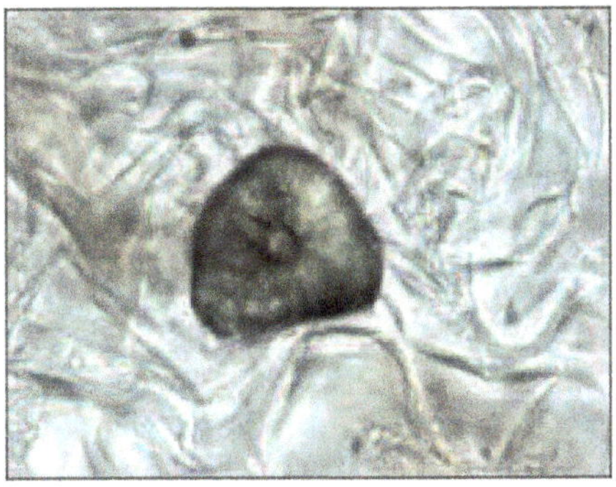

Fresh mount of a sclereid

Long, tapered sclereids supporting a leaf edge in Dionysia kossinskyi

They are small bundles of sclerenchyma tissue in plants that form durable layers, such as the cores of apples and the gritty texture of pears (Pyrus communis). Sclereids are variable in shape. The cells can be isodiametric, prosenchymatic, forked or elaborately branched. They can be grouped into bundles, can form complete tubes located at the periphery or can occur as single cells or small groups of cells within parenchyma tissues. But compared with most fibres, sclereids are relatively short. Characteristic examples are brachysclereids or the stone cells (called stone cells because of their hardness) of pears and quinces (Cydonia oblonga) and those of the shoot of the wax plant (Hoya carnosa). The cell walls fill nearly all the cell's volume. A layering of the walls and the existence of branched pits is clearly visible. Branched pits such as these are called ramiform pits. The shell of many seeds like those of nuts as well as the stones of drupes like cherries and plums are made up from sclereids.

These structures are used to protect other cells.

Phragmoplast

The **phragmoplast** is a plant cell specific structure that forms during late cytokinesis. It serves as a scaffold for cell plate assembly and subsequent formation of a new cell wall separating the two daughter cells.

Phragmoplast and cell plate formation in a plant cell during cytokinesis. Left side: Phragmoplast forms and cell plate starts to assemble in the center of the cell. Towards the right: Phragmoplast enlarges in a donut-shape towards the outside of the cell, leaving behind mature cell plate in the center. The cell plate will transform into the new cell wall once cytokinesis is complete.

The phragmoplast is a complex assembly of microtubules (MTs), microfilaments (MFs), and endoplasmic reticulum (ER) elements, that assemble in two opposing sets perpendicular to the plane of the future cell plate during anaphase and telophase. It is initially barrel-shaped and forms from the mitotic spindle between the two daughter nuclei while nuclear envelopes reassemble around them. The cell plate initially forms as a disc between the two halves of the phragmoplast structure. While new cell plate material is added to the edges of the growing plate, the phragmoplast microtubules disappear in the center and regenerate at the edges of the growing cell plate. The two structures grow outwards until they reach the outer wall of the dividing cell. If a phragmosome was present in the cell, the phragmoplast and cell plate will grow through the space occupied by the phragmosome. They will reach the parent cell wall exactly at the position formerly occupied by the preprophase band.

The microtubules and actin filaments within the phragmoplast serve to guide vesicles with cell wall material to the growing cell plate. Actin filaments are also possibly involved in guiding the phragmoplast to the site of the former preprophase band location at the parent cell wall. While the cell plate is growing, segments of smooth endoplasmic reticulum are trapped within it, later forming the plasmodesmata connecting the two daughter cells.

The phragmoplast can only be observed in Embryophytes, that is the bryophytes and vascular plants, and a few advanced green algae, specifically Coleochaete in the Division Charophyta. Some algae use another type of microtubule array, a phycoplast, during cytokinesis.

REFERENCES

1. Venes, Donald (2001). Taber's Cyclopedic Medical Dictionary (Twentieth Edition), (F.A. Davis Company, Philadelphia), p. 2287 ISBN 0-9762548-3-2.
2. Heide N. Schulz-Vogt (2006). Vacuoles. Microbiology Monographs. 1. doi:10.1007/3-540-33774-1_10. ISBN 3-540-26205-9.
3. Brooker, Robert J, et al. (2007). Biology (First Edition), (McGraw-Hill, New York), p. 79 ISBN 0-07-326807-0.
4. http://pcp.oxfordjournals.org/content/50/1/151.full.pdf
5. Matile, Phillipe (1993) Chapter 18: Vacuoles, discovery of lysosomal origin in Discoveries in Plant Biology: v. 1 (World Scientific Publishing Co Pte Ltd)

6. Thomas Boller. Plantbiology.unibas.ch. Retrieved on 2011-09-02.
7. Jezbera Jan; Karel Hornak; Karel Simek (2005). "Food selection by bacterivorous protists: insight from the analysis of the food vacuole by means of fluorescence in situ hybridization". FEMS Microbiology Ecology. 52 (3): 351–363. doi:10.1016/j.femsec.2004.12.001. PMID 16329920.
8. Becker B (2007). "Function and evolution of the vacuolar compartment in green algae and land plants (Viridiplantae)." (PDF). International Review of Cytology. 264: 1–24. doi:10.1016/S0074-7696(07)64001-7. PMID 17964920.
9. Spallanzani, L. [1776]. 1920. Observations et expériences faites sur les Animalicules des Infusions. L' École Polytechnique, Paris
10. Dujardin, F. 1841. Histoire naturelle des zoophytes: Infusoires. Librairie Encyclopédique de Roret, Paris.
11. Schleiden, M. J. 1842. Grundzüge der wissenschaftlichen Botanik. W. Engelmann, Leipzig
12. Wayne, R. 2009. Plant Cell Biology: From Astronomy to Zoology. Amsterdam: Elsevier/Academic Press, p. 101.
13. de Vries, H. 1885. Plasmolytische Studien über die Wand der Vakuolen. Jahrb. wiss. Bot. 16, 465–598
14. Kalanetra KM, Huston SL, Nelson DC (December 2004). "Novel, attached, sulfur-oxidizing bacteria at shallow hydrothermal vents possess vacuoles not involved in respiratory nitrate accumulation". Appl. Environ. Microbiol. 70 (12): 7487–96. doi:10.1128/AEM.70.12.7487-7496.2004. PMC 535177Freely accessible. PMID 15574952.
15. Walsby AE (1969). "The Permeability of Blue-Green Algal Gas-Vacuole Membranes to Gas". Proceedings of the Royal Society of London. Series B, Biological Sciences. 173 (1031): 235–255. doi:10.1098/rspb.1969.0049. ISSN 0080-4649. OCLC 479422015.
16. Alberts, Bruce, Johnson, Alexander, Lewis, Julian, Raff, Martin, Roberts, Keith, and Walter, Peter (2008). Molecular Biology of the Cell (Fifth Edition), (Garland Science, New York), p. 781 ISBN 978-0-8153-4111-6.
17. Li, WING-YEN; FUK-LING WONG; SAU-NA TSAI; TSUI-HUNG PHANG; GUIHUA SHAO; HON-MING LAM (June 2006). "Tonoplast-located GmCLC1 and GmNHX1 from soybean enhance NaCl tolerance in transgenic bright yellow (BY)-2 cells". Plant, Cell and Environment. 29 (6): 1122–1137. doi:10.1111/j.1365-3040.2005.01487.x. PMID 17080938. Retrieved 23 October 2011.
18. Lincoln Taiz and Eduardo Zeiger Plant Physiology 3rd Edition SINAUER 2002 p.13 and 14 ISBN 0-87893-856-7
19. D J Klionsky; P K Herman & S D Emr (1990). "The fungal vacuole: composition, function, and biogenesis". Microbiol Rev. 54 (3): 266–292. PMC 372777Freely accessible. PMID 2215422.
20. Plant cells vs. Animal cells. Biology-Online.org
21. William F. Ganong, MD (2003). Review of medical physiology (21st ed.).

22. Reggiori F (2006). "Membrane Origin for Autophagy". Current Topics in Developmental Biology. 74: 1–30. doi:10.1016/S0070-2153(06)74001-7. PMID 16860663.

23. Leigh A. Knodler; Olivia Steele-Mortimer (2003). "Taking Possession: Biogenesis of the Salmonella-Containing Vacuole". Traffic. 4 (9): 587–599. doi:10.1034/j.1600-0854.2003.00118.x. PMID 12911813.

CHAPTER 4

Flower Reproduction

Chapter Contents

- Abc Model of Flower Development
- Floral Transition
- Formation of the Floral Meristem or the Inflorescence
- Floral Architecture
- Inflorescence
- Flower
- Evolution
- Perianth
- Sporophyll
- Archegonium
- Stamen
- Sporophyte
- Fruit

Reproduction in flowering plants begins with pollination, the transfer of pollen from anther to stigma on the same flower or to the stigma of another flower on the same plant (self-pollination), or from anther on one plant to the stigma of another plant (cross-pollination). Once the pollen grain lodges on the stigma, a pollen tube grows from the pollen grain to an ovule. Two sperm nuclei then pass through the pollen tube. One of them unites with the egg nucleus and produces a zygote. The other sperm nucleus unites with two polar nuclei to produce an endosperm nucleus. The fertilized ovule develops into a seed.

ABC MODEL OF FLOWER DEVELOPMENT

ABC model of flower development is a scientific model of the process by which flowering plants produce a pattern of gene expression in meristems that leads to the appearance of an organ oriented towards sexual reproduction, a flower. There are three physiological developments that must occur in order for this to take place: firstly, the plant must pass from sexual immaturity into a sexually mature state (i.e. a transition towards flowering); secondly, the transformation of the apical meristem's function from a vegetative meristem into a floral meristem or inflorescence; and finally the growth of the flower's individual organs. The latter phase has been modelled using the **ABC model**, which aims to describe the biological basis of the process from the perspective of molecular and developmental genetics.

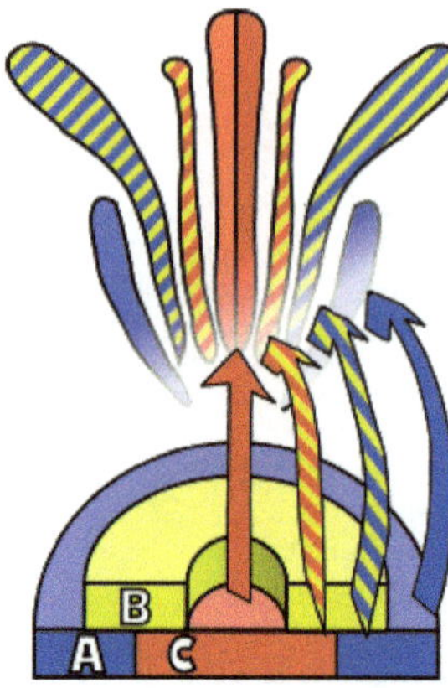

A diagram illustrating the ABC model in Arabidopsis. Class A genes (blue) affect sepals and petals, class B genes (yellow) affect petals and stamens, class C genes (red) affect stamens and carpels. In two specific whorls of the floral meristem, each class of organ identity genes is switched on.

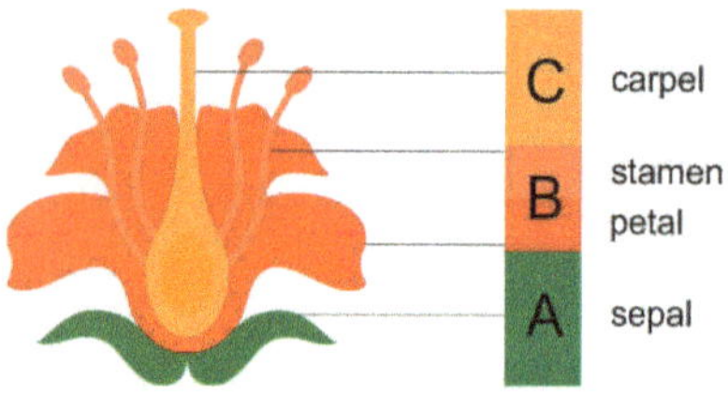

A diagram illustrating the ABC model. Class A genes affect sepals and petals, class B genes affect petals and stamens, class C genes affect stamens and carpels. In two specific whorls of the floral meristem, each class of organ identity genes is switched on.

An external stimulus is required in order to trigger the differentiation of the meristem into a flower meristem. This stimulus will activate mitotic cell division in the meristem, particularly on its sides where new primordia are formed. This same stimulus will also cause the meristem to follow a developmental pattern that will lead to the growth of floral meristems as opposed to vegetative meristems. The main difference between these two types of meristem, apart from the obvious disparity between the objective organ, is the verticillate (or whorled) phyllotaxis, that is, the absence of stem elongation among the successive whorls or verticils of the primordium. These verticils follow an acropetal development, giving rise to sepals, petals, stamens and carpels. Another difference from vegetative axillary meristems is that the floral meristem is "determined", which means that, once differentiated, its cells will no longer divide.

The identity of the organs present in the four floral verticils is a consequence of the interaction of at least three types of gene products, each with distinct functions. According to the ABC model, functions A and C are required in order to determine the identity of the verticils of the perianth and the reproductive verticils, respectively. These functions are exclusive and the absence of one of them means that the other will determine the identity of all the floral verticils. The B function allows the differentiation of petals from sepals in the secondary verticil, as well as the differentiation of the stamen from the carpel on the tertiary verticil. Goethe's "foliar theory" was formulated in the 18th century and it suggests that the constituent parts of a flower are structurally modified leaves, which are functionally specialized for reproduction or protection. The theory was first published in 1790 in the essay "Metamorphosis of Plants" ("*Versuch die Metamorphose der Pflanzen zu erklaren*"). where Goethe wrote:

"...we may equally well say that a stamen is a contracted petal, as that a petal is a stamen in a state of expansion; or that a sepal is a contracted stem leaf approaching a certain stage of refinement, as that a stem leaf is a sepal expanded by the influx of cruder saps".

FLORAL TRANSITION

The transition from the vegetative phase to a reproductive phase involves a dramatic change in the plant's vital cycle, perhaps the most important one, as the process must be carried out correctly in order to guarantee that the plant produces descendents. This transition is characterised by the induction and development of the meristem of the inflorescence, which will produce a collection of flowers or one flower, where only one is produced. This morphogenetic change contains both endogenous and exogenous elements: For example, in order for the change to be initiated the plant must have a certain number of leaves and contain a certain level of total biomass. Certain environmental conditions are also required such as a characteristic photoperiod. Plant hormones play an important part in the process, with the gibberellins having a particularly important role.

There are many signals that regulate the molecular biology of the process. The following three genes in *Arabidopsis thaliana* possess both common and independent functions in floral transition: *FLOWERING LOCUS T* (*FT*), *LEAFY* (*LFY*), *SUPPRESOR OF OVEREXPRESSION OF CONSTANS1* (*SOC1*, also called *AGAMOUS-LIKE20*). *SOC1* is a MADS-box-type gene, which integrates responses to photoperiod, vernalization and gibberellins.

FORMATION OF THE FLORAL MERISTEM OR THE INFLORESCENCE

The meristem can be defined as the tissue or group of plant tissues that contain undifferentiated stem cells, which are capable of producing any type of cell tissue. Their maintenance and development, both in the vegetative meristem or the meristem of the inflorescence is controlled by genetic cell fate determination mechanisms. This means that a number of genes will directly regulate, for example, the maintenance of the stem cell's characteristics (gene *WUSCHEL* or *WUS*), and others will act via negative feedback mechanisms in order to inhibit a characteristic (gene *CLAVATA* or *CLV*). In this way both mechanisms give rise to a feedback loop, which along with other elements lend a great deal of robustness to the system. Along with the *WUS* gene the *SHOOTMERISTEMLESS* (*STM*) gene also represses the differentiation of the meristematic dome. This gene acts by inhibiting the possible differentiation of the stem cells but still allows cell division in the daughter cells, which, had they been allowed to differentiate, would have given rise to distinct organs.

FLORAL ARCHITECTURE

A flower's anatomy, as defined by the presence of a series of organs (sepals, petals, stamens and carpels) positioned according to a given pattern, facilitate sexual reproduction in flowering plants. The flower arises from the activity of three classes of genes, which regulate floral development: genes which regulate the identity of the meristem, the identity of the flower organ and finally cadastral genes.

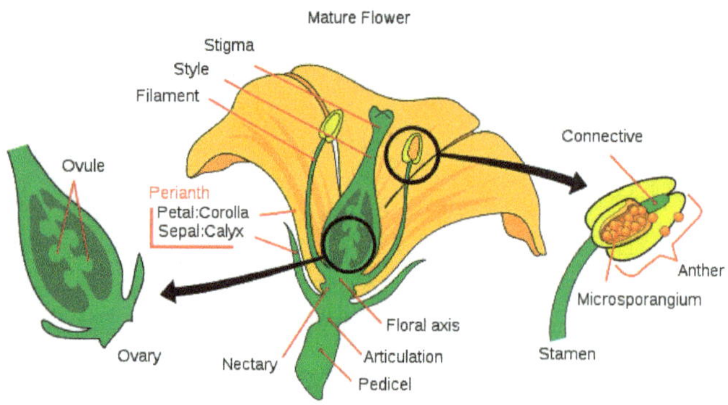

Anatomy of a flower.

- Meristem identity genes. Code for the transcription factors required to initiate the induction of the identity genes. They are positive regulators of organ identity during floral development.
- Organ identity genes. Directly control organ identity and also code for transcription factors that control the expression of other genes, whose products are implicated in the formation or function of the distinct organs of the flower.
- Cadastral genes. Act as spatial regulators for the organ identity genes by defining boundaries for their expression. In this way they control the extent to which genes interact thereby regulating whether they act in the same place at the same time.

The ABC Model

The ABC model of flower development was first formulated by George Haughn and Chris Somerville in 1988. It was first used as a model to describe the collection of genetic mechanisms that establish floral organ identity in the Rosids, as exemplified by *Arabidopsis thaliana*, and the Asterids, as demonstrated by *Antirrhinum majus*. Both species have four verticils (sepals, petals, stamens and carpels), which are defined by the differential expression of a number of homeotic genes present in each verticil. This means that the sepals are solely characterized by the expression of A genes, while the petals are characterized by the co-expression of A and B genes. The B and C genes establish the identity of the stamens and the carpels only require C genes to be active. It should be noted that type A and C genes are reciprocally antagonistic.

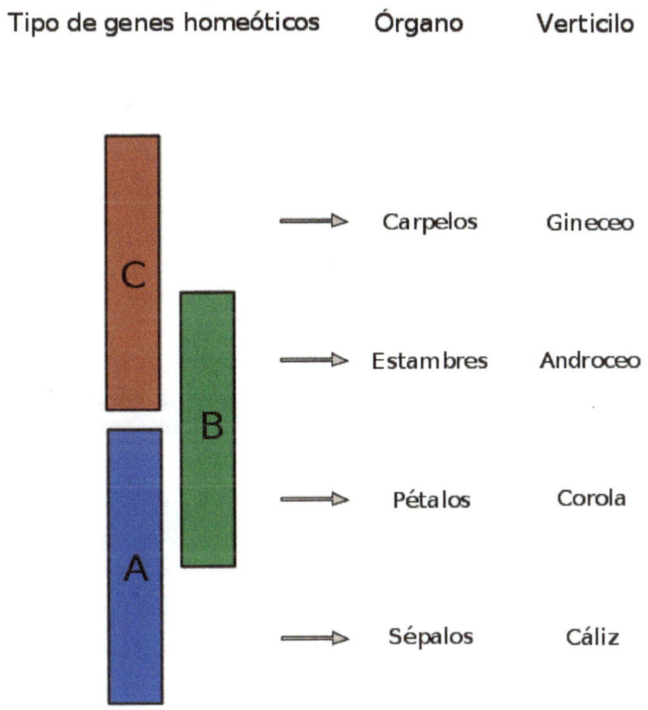

Graphic representation of the ABC model. The single or additive expression of the homeotic genes in the left hand column have repercussions for the development of the organs in the central column and determine the nature of the whorl, on the right.

The fact that these homeotic genes determine an organ's identity becomes evident when a gene that represents a particular function, for example the A gene, is not expressed. In *Arabidopsis* this loss results in a flower which is composed of one verticil of carpels, another containing stamens and another of carpels. This method for studying gene function uses reverse genetics techniques to produce transgenic plants that contain a mechanism for gene silencing through RNA interference. In other studies, using forward genetics techniques such as genetic mapping, it is the analysis of the phenotypes of flowers with structural anomalies that leads to the cloning of the gene of interest. The flowers may possess a non-functional or over expressed allele for the gene being studied.

The existence of two supplementary functions, D and E, have also been proposed in addition to the A, B and C functions already discussed. Function D specifies the identity of the ovule, as a separate reproductive function from the development of the carpels, which occurs after their determination. Function E relates to a physiological requirement that is a characteristic of all floral verticils, although, it was initially described as necessary for the development of the three innermost verticils (Function E *sensu stricto*). However, its broader definition (*sensu lato*) suggests that it is required in the four verticils.Therefore, when Function D is lost the structure of the ovules becomes similar to that of leaves and when Function E is lost *sensu stricto*, the floral organs of the three outer most verticils are transformed into sepals, while on losing Function E *sensu lato*, all the verticils are similar to leaves. It is interesting to note that the gene products of genes with D and E functions are also MADS-box genes.

Genetic Analysis

The methodology for studying flower development involves two steps. Firstly, the identification of the exact genes required for determining the identity of the floral meristem. In *A. thaliana* these include APETALA1 (*AP1*) and LEAFY (*LFY*). Secondly, genetic analysis is carried out on the aberrant phenotypes for the relative characteristics of the flowers, which allows the characterization of the homeotic genes implicated in the process.

Flower of *A. thaliana*.

Flowers of *A. majus*.

Flowers of *Petunia hybrid*.

Analysis of Mutants

There are a great many mutations that affect floral morphology, although the analysis of these mutants is a recent development. Supporting evidence for the existence of these mutations comes from the fact that a large number affect the identity of floral organs. For example, some organs develop in a location where others should develop. This is called homeotic mutation, which is analogous to HOX gene mutations found in *Drosophila*. In *Arabidopsis* and *Antirrhinum*, the two taxa on which models are based, these mutations always affect adjacent verticils. This allows the characterization of three classes of mutation, according to which verticils are affected:

- Mutations in type A genes, these mutations affect the calyx and corolla, which are the outermost verticils. In these mutants, such as APETALA2 in *A. thaliana*, carpels develop instead of sepals and stamen in place of petals. This means that, the verticils of the perianth are transformed into reproductive verticils.
- Mutations in type B genes, these mutations affect the corolla and the stamen, which are the intermediate verticils. Two mutations have been found in *A. thaliana*, APETALA3 and PISTILLATA, which cause development of sepals instead of petals and carpels in the place of stamen.
- Mutations in type C genes, these mutations affect the reproductive verticils, namely the stamen and the carpels. The *A. thaliana* mutant of this type is called AGAMOUS, it possesses a phenotype containing petals instead of stamen and sepals instead of carpels.

Techniques for Detecting Differential Expression

Cloning studies have been carried out on DNA in the genes associated with the affected homeotic functions in the mutants discussed above. These studies used serial analysis of gene expression throughout floral development to show patterns of tissue expression, which, in general, correspond with the predictions of the ABC model.

The nature of these genes corresponds to that of transcription factors, which, as expected, have analogous structures to a group of factors contained in yeasts and animal cells. This group is called MADS, which is an acronym for the different factors contained in the group. These MADS factors have been detected in all the vegetable species studied, although the involvement of other elements involved in the regulation of gene expression cannot be discounted.

Genes Exhibiting type-A Function

In *A. thaliana*, function A is mainly represented by two genes *APETALA1* (*AP1)* and *APETALA2* (*AP2*) *AP1* is a MADS-box type gene, while *AP2* belongs to the family of genes that contains AP2, which it gives its name to and which consists of transcription factors that are only found in plants. AP2 has also been shown to complex with the co-repressor TOPLESS (TPL) in developing floral buds to repress the C-class gene *AGAMOUS* (*AG*). However, *AP2* is not expressed in the shoot apical meristem (SAM), which contains the latent stem cell population throughout the adult life of *Arabidopsis*, and so it is speculated that TPL works with some other A-class gene in the SAM to repress *AG.AP1* functions as a type A gene, both in controlling the identity of sepals and petals, and it also acts in the floral meristem. *AP2* not only functions in the first two verticils, but also in the remaining two, in developing ovules and even in leaves. It is also likely that post-transcriptional regulation exists, which controls it's A function, or even that it has other purposes in the determination of organ identity independent of that mentioned here.

In *Antirrhinum*, the orthologous gene to *AP1* is *SQUAMOSA* (*SQUA*), which also has a particular impact on the floral meristem. The homologs for *AP2* are *LIPLESS1* (*LIP1*) and *LIPLESS2* (*LIP2*), which have a redundant function and are of special interest in the development of sepals, petals and ovules.

A total of three genes have been isolated from *Petunia hybrida* that are similar to *AP2*: *P. hybrida APETALA2A* (*PhAP2A*), *PhAP2B* and *PhAP2C*. *PhAP2A* is, to a large degree, homologous with the *AP2* gene of *Arabidopsis*, both in its sequence and in its expression pattern, which suggests that the two genes are orthologs. The proteins *PhAP2B* and *PhAP2C*, on the other hand, are slightly different, even though they belong to the family of transcription factors that are similar to *AP2*. In addition they are expressed in different ways, although they are very similar in comparison with *PhAP2A*. In fact, the mutants for these genes do not show the usual phenotype, that of the null alleles of A genes. A true A-function gene has not been found in Petunia; though a part of the A-function (the inhibition of the C in the outer two whorls) has been largely attributed to miRNA169 (colloquially called BLIND)[ref].

Genes Exhibiting type-B Function

In *A. thaliana* the type-B function mainly arises from two genes, *APETALA3* (*AP3*) and *PISTILLATA* (*PI*), both of which are MADS-box genes. A mutation of either of these genes causes the homeotic conversion of petals into sepals and of stamens into carpels. This also occurs in its orthologs in *A. majus*, which are DEFICIENS (*DEF*) and *GLOBOSA* (*GLO*) respectively. For both species the active form of binding with DNA is that derived from the heterodimer: AP3 and PI, or DEF and GLO, dimerize. This is the form in which they are able to function.

The *GLO/PI* lines that have been duplicated in *Petunia* contain *P. hybrida GLOBOSA1* (*PhGLO1*, also called *FBP1*) and also *PhGLO2* (also called *PMADS2* or *FBP3*). For the functional elements equivalent to *AP3/DEF* in *Petunia* there is both a gene that possesses a relatively similar sequence, called *PhDEF* and there is also an atypical B function gene called PhTM6. Phylogenetic studies have placed the first three within the «euAP3» lineage, while PhTM6 belongs to that of «paleoAP3». It is worth pointing out that, in terms of evolutionary history, the appearance of the euAP3 line seems to be related with the emergence of dicotyledons, as representatives of euAP3-type B function genes are present in dicotyledons while paleoAP3 genes are present in monocotyledons and basal angiosperms, among others.

As discussed above, the floral organs of eudicotyledonous angiosperms are arranged in 4 different verticils, containing the sepals, petals, stamen and carpels. The ABC model states that the identity of these organs is determined by the homeotic genes A, A+B, B+C and C, respectively. In contrast with the sepal and petal verticils of the eudicots, the perigone of many plants of the Liliaceae family have two nearly identical external petaloid verticils (the tepals). In order to explain the floral morphology of the Liliaceae, van Tunen *et al.* proposed a modified ABC model in 1993. This model suggests that class B genes are not only expressed in verticils 2 and 3, but also in 1. It therefore follows that the organs of verticils 1 and 2 express class A and B genes and this is how they have a petaloid structure. This theoretical model has been experimentally proven through the cloning and characterization of homologs of the *Antirrhinum* genes *GLOBOSA* and *DEFICIENS* in a Liliaceae, the tulip *Tulipa gesneriana*. These genes are expressed in verticils 1,2 and 3. The homologs *GLOBOSA* and *DEFICIENS* have also been isolated and characterized in *Agapanthus praecox* ssp. *orientalis* (Agapanthaceae), which is phylogenetically distant from the model organisms. In this study the genes were called *ApGLO* and *ApDEF*, respectively. Both contain open reading frames that code for proteins with 210 to 214 amino acids. Phylogenetic analysis of these sequences indicated that they belong to B gene family of the monocotyledons. In situ hybridization studies revealed that both sequences are expressed in verticil 1 as well as in 2 and 3. When taken together, these observations show that the floral development mechanism of *Agapanthus* also follows the modified ABC model.

Genes Exhibiting type-C Function

In *A. thaliana*, the C function is derived from one MADS-box type gene called *AGAMOUS* (*AG*), which intervenes both in the establishment of stamen and carpel identity as well as in the determination of the floral meristem. Therefore, the *AG* mutants are devoid of androecium and gynoecium and they have petals and sepals in their place. In addition, the growth in the centre of the flower is undifferentiated, therefore the petals and sepals grow in repetitive verticils.

The *PLENA* (*PLE*) gene is present in *A. majus*, in place of the *AG* gene, although it is not an ortholog. However, the *FARINELLI* (*FAR*) gene is an ortholog, which is specific to the development of the anthers and the maturation of pollen.

In *Petunia*, *Antirrhinum* and in maize the C function is controlled by a number of genes that act in the same manner. The genes that are closer homologs of *AG* in *Petunia* are *pMADS3* and *floral-binding protein 6* (*FBP6*).

Genes Exhibiting type-D and E Functions

The D function genes were discovered in 1995. These genes are MADS-box proteins and they have a function that is distinct from those previously described, although they have a certain homology with C function genes. These genes are called *FLORAL BINDING PROTEIN7* (*FBP7*) and *FLORAL BINDING PROTEIN1L* (*FBP1l*). It was found that, in *Petunia*, they are involved in the development of the ovule. Equivalent genes were later found in *Arabidopsis*, where they are also involved in controlling the development of carpels and the ovule and even with structures related to seed dispersal.

The appearance of interesting phenotypes in RNA interference studies in *Petunia* and tomato led, in 1994, to the definition of a new type of function in the floral development model. The E function was initially thought to be only involved in the development of the three innermost verticils, however, subsequent work found that its expression was required in all the floral verticils.

INFLORESCENCE

An **inflorescence** is a group or cluster of flowers arranged on a stem that is composed of a main branch or a complicated arrangement of branches. Morphologically, it is the part of the shoot of seed plants where flowers are formed and which is accordingly modified. The modifications can involve the length and the nature of the internodes and the phyllotaxis, as well as variations in the proportions, compressions, swellings, adnations, connations and reduction of main and secondary axes. Inflorescence can also be defined as the reproductive portion of a plant that bears a cluster of flowers in a specific pattern.

Aloe hereroensis, showing inflorescence with branched peduncle

The stem holding the whole inflorescence is called a peduncle and the major axis (incorrectly referred to as the main stem) holding the flowers or more branches within the inflorescence is called the **rachis**. The stalk of each single flower is called a pedicel. A flower that is not part of an inflorescence is called a solitary flower and its stalk is also referred to as a peduncle. Any flower in an inflorescence may be referred to as a **floret**, especially when the individual flowers are particularly small and borne in a tight cluster, such as in a pseudanthium. The fruiting stage of an inflorescence is known as an infructescence.

Inflorescences may be simple (single) or complex (panicle). The rachis may be one of several types, including Single, Composite, Umbel, Spike or Raceme.

General Characteristics

Inflorescences are described by many different characteristics including how the flowers are arranged on the peduncle, the blooming order of the flowers and how different clusters of flowers are grouped within it. These terms are general representations as plants in nature can have a combination of types.

Bracts

Inflorescences usually have modified shoots foliage different from the vegetative part of the plant. Considering the broadest meaning of the term, any leaf associated with an inflorescence is called a **bract**. A bract is usually located at the node where the main stem of the inflorescence forms, joined to the main stem of the plant, but other bracts can exist within the inflorescence itself. They serve a variety of functions which include attracting pollinators and protecting young flowers. According to the presence or absence of bracts and their characteristics we can distinguish:

- **Ebracteate inflorescences**: No bracts in the inflorescence.
- **Bracteate inflorescences**: The bracts in the inflorescence are very specialised, sometimes reduced to small scales, divided or dissected.
- **Leafy inflorescences**: Though often reduced in size, the bracts are unspecialised and look like the typical leaves of the plant, so that the term **flowering stem** is usually applied instead of inflorescence. This use is not technically correct, as, despite their 'normal' appearance, these *leaves* are considered, in fact, *bracts*, so that 'leafy inflorescence' is preferable.
- **Leafy-bracted inflorescences**: Intermediate between bracteate and leafy inflorescence.

If many bracts are present and they are strictly connected to the stem, like in the family Asteraceae, the bracts might collectively be called an **involucre**. If the inflorescence has a second unit of bracts further up the stem, they might be called an **involucel**.

Ebracteate inflorescence.

Ebracteate of *Wisteria sinensis*

Bracteate inflorescence.

Bracteate inflorescence of *Pedicularis verticillata*.

Leafy-bracted inflorescence.

Leafy-bracted inflorescence of *Rhinanthus angustifolius*.

Leafy inflorescence.

Leafy inflorescence of *Aristolochia clematitis*.

Terminal Flower

Plant organs can grow according to two different schemes, namely monopodial or **racemose** and sympodial or **cymose**. In inflorescences these two different growth patterns are called indeterminate and determinate respectively, and indicate whether a terminal flower is formed and where flowering starts within the inflorescence.

- **Indeterminate inflorescence**: Monopodial (racemose) growth. The terminal bud keeps growing and forming lateral flowers. A terminal flower is never formed.
- **Determinate inflorescence**: Sympodial (cymose) growth. The terminal bud forms a terminal flower and then dies out. Other flowers then grow from lateral buds.

Indeterminate and determinate inflorescences are sometimes referred to as **open** and **closed** inflorescences respectively.

In an **indeterminate inflorescence** there is no true terminal flower and the stem usually has a rudimentary end. In many cases the last true flower formed by the terminal bud (**subterminal** flower) straightens up, appearing to be a terminal flower. Often a vestige of the terminal bud may be noticed higher on the stem.

Indeterminate inflorescence with a perfect acropetal maturation.

Indeterminate inflorescence with an acropetal maturation and lateral flower buds.

Indeterminate inflorescence with the subterminal flower to simulate the terminal one (vestige present)

In **determinate inflorescences** the terminal flower is usually the first to mature (precursive development), while the others tend to mature starting from the bottom of the stem. This pattern is called **acropetal** maturation. When flowers start to mature from the top of the stem, maturation is **basipetal**, while when the central mature first, **divergent**.

Determinate inflorescence with acropetal maturation

Determinate inflorescence with basipetal maturation

Determinate inflorescence with divergent maturation

Phyllotaxis

As with leaves, flowers can be arranged on the stem according to many different patterns. See 'Phyllotaxis' for in-depth descriptions.

Alternate flowers

Opposite flowers

Metatopy

Metatopy is the placement of organs out of their normally expected position: typically metatopy occurs in inflorescences when unequal growth rates alter different areas of the axis and the organs attached to the axis.

When a single or a cluster of flower(s) is located at the axil of a bract, the location of the bract in relation to the stem holding the flower(s) is indicated by the use of different terms and may be a useful diagnostic indicator.

Typical placement of bracts include:

- Some plants have bracts that subtend the inflorescence, where the flowers are on branched stalks; the bracts are not connected to the stalks holding the flowers, but are adnate or attached to the main stem (Adnate describes the fusing together of different unrelated parts. When the parts fused together are the same, they are connately joined.)
- Other plants have the bracts subtend the pedicel or peduncle of single flowers.

Metatopic placement of bracts include:

- When the bract is attached to the stem holding the flower (the pedicel or peduncle), it is said to be **recaulescent**; sometimes these bracts or bracteoles are highly modified and appear to be appendages of the flower calyx. Recaulescences is the fusion of the subtending leaf with the stem holding the bud or the bud itself, thus the leaf or bract is adnate to the stem of flower.
- When the formation of the bud is shifted up the stem distinctly above the subtending leaf, it is described as **concaulescent**.

Flower and subtending bract

Lilium martagon (flower and subtending bract)

Concaulescence

Solanum lycopersicum (concaulescence)

Recaulescence

Tilia cordata (recaulescence)

Organization

There is no general consensus in defining the different inflorescences. The following is based on Focko Weberling's *Morphologie der Blüten und der Blütenstände* (Stuttgart, 1981). The main groups of inflorescences are distinguished by branching. Within these groups, the most important characteristics are the intersection of the axes and different variations of the model. They may contain many flowers (**pluriflor**) or a few (**pauciflor**). Inflorescences can be **simple** or **compound**.

Simple Inflorescences

Indeterminate or Racemose

Indeterminate simple inflorescences are generally called **racemose**. The main kind of racemose inflorescence is the **raceme**. The other kind of racemose inflorescences can all be derived from this one by dilation, compression, swelling or reduction of the different axes. Some passage forms between the obvious ones are commonly admitted.

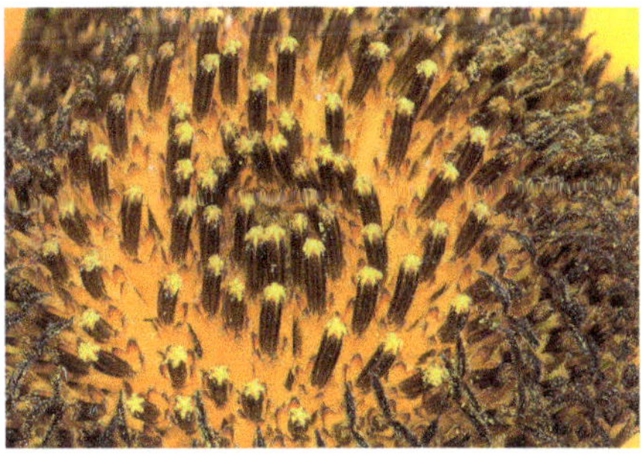

Inflorescence of sessile disc florets forming the capitulum

- A **raceme** is an unbranched, indeterminate inflorescence with pedicellate (having short floral stalks) flowers along the axis.
- A **spike** is a type of raceme with flowers that do not have a pedicel.
- A racemose **corymb** is an unbranched, indeterminate inflorescence that is flat-topped or convex due to their outer pedicels which are progressively longer than inner ones.
- An **umbel** is a type of raceme with a short axis and multiple floral pedicels of equal length that appear to arise from a common point.

- A **spadix** is a spike of flowers densely arranged around it, enclosed or accompanied by a highly specialised bract called a **spathe**. It is characteristic of the Araceae family.
- A **flower head** or **capitulum** is a very contracted raceme in which the single sessile flowers share are borne on an enlarged stem. It is characteristic of Dipsacaceae.
- A **catkin** or **ament** is a scaly, generally drooping spike or raceme. Cymose or other complex inflorescences that are superficially similar are also generally called thus.

Raceme

Epilobium angustifolium

Spike

Plantago media (spike)

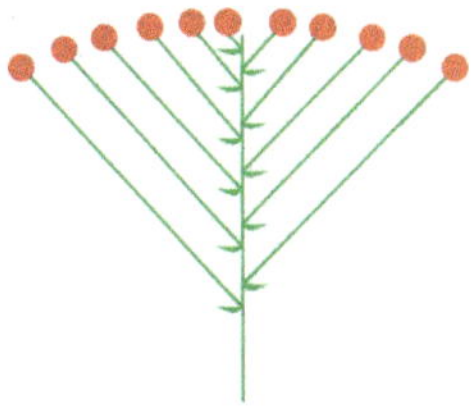

Racemose corymb

Iberis umbellata (racemose corymb)

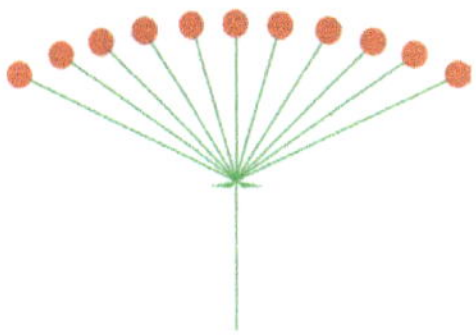

Umbel

Astrantia minor (umbel)

Spadix

Arum maculatum (spadix)

Head (round)

Dipsacus fullonum (head)

Catkin (racemose or spicate)

Alnus incana (ament)

Determinate or Cymose

Determinate simple inflorescences are generally called **cymose**. The main kind of cymose inflorescence is the **cyme** (pronounced 'saim', from the Latin *cyma* in the sense 'cabbage sprout', from Greek *kuma* 'anything swollen'). Cymes are further divided according to this scheme:

- Only one secondary axis: **monochasium**
- Secondary buds always develop on the same side of the stem: **helicoid cyme** or **bostryx**
- The successive pedicels are aligned on the same plane: **drepanium**
- Secondary buds develop alternately on the stem : **scorpioid cyme**
- The successive pedicels are arranged in a sort of spiral: **cincinnus** (characteristic of the Boraginaceae and Commelinaceae)
- The successive pedicels follow a zig-zag path on the same plane: **rhipidium** (many Iridaceae)
- Two secondary axes: **dichasial cyme**
    - Secondary axis still dichasial: **dichasium** (characteristic of Caryophyllaceae)
    - Secondary axis monochasia: **double scorpioid cyme** or **double helicoid cyme**
- More than two secondary axes: **pleiochasium**

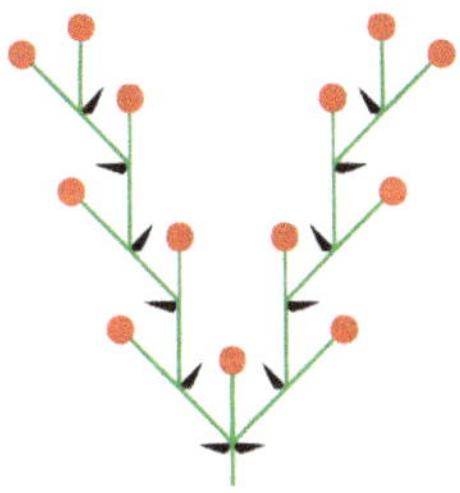

Double cyme

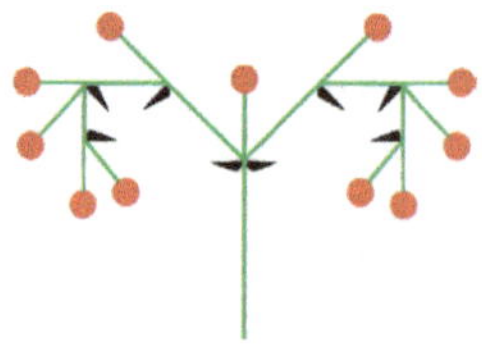

Double cyme

Bostryx (lateral and top view)

Hypericum perforatum (bostryx)

Drepanium (lateral and top view)

Gladiolus imbricatus (drepanium)

Cincinnus (lateral and top view)

Symphytum officinale (cincinnus)

Rhipidium (lateral and top view)

Canna sp. (rhipidium)

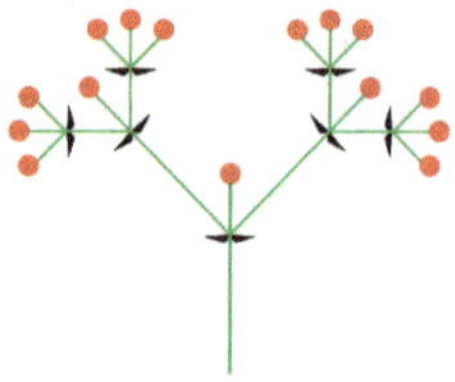

Dichasium

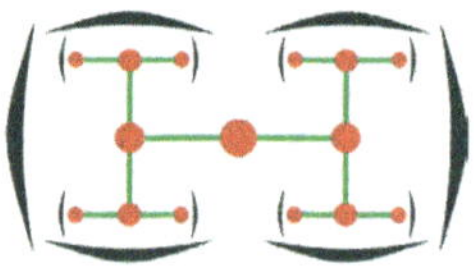

Dichasium, top view

Silene dioica (dichasium)

A cyme can also be so compressed that it looks like an umbel. Strictly speaking this kind of inflorescence could be called **umbelliform cyme**, although it is normally called simply 'umbel'.

Another kind of definite simple inflorescence is the raceme-like cyme or **botryoid**; that is as a raceme with a terminal flower and is usually improperly called 'raceme'.

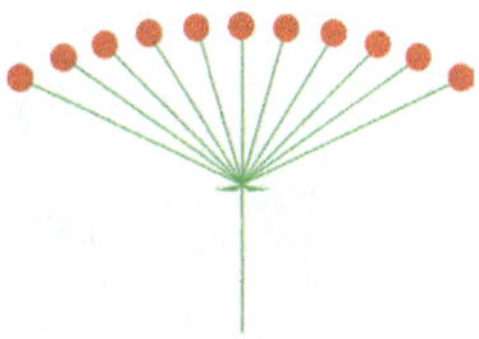

Umbelliform cyme

Pelargonium zonale (umbelliform cyme)

Botryoid

Berberis vernae (botryoid)

A reduced raceme or cyme that grows in the axil of a bract is called a **fascicle**. A **verticillaster** is a fascicle with the structure of a dichasium; it is common among the Lamiaceae. Many verticillasters with reduced bracts can form a spicate (spike-like) inflorescence that is commonly called a **spike**.

Gentiana lutea (fascicles)

Lamium orvala (verticillaster)

Mentha longifolia ('spike')

Compound Inflorescences

Simple inflorescences are the basis for compound inflorescences or synflorescences. The single flowers are there replaced by a simple inflorescence, which can be both a racemose or a cymose one. Compound inflorescences are composed of branched stems and can involve complicated arrangements that are difficult to trace back to the main branch.

A kind of compound inflorescence is the double inflorescence, in which the basic structure is repeated in the place of single florets. For example a double raceme is a raceme in which the single flowers are replaced by other simple racemes; the same structure can be repeated to form triple or more complex structures.

Compound raceme inflorescences can either end with a final raceme (homoeothetic), or not (heterothetic). A compound raceme is often called a panicle. Note that this definition is very different from that given by Weberling. Compound umbels are umbels in which the single flowers are replaced by many smaller umbels called umbellets. The stem attaching the side umbellets to the main stem is called a ray.

Homeothetic compound raceme

Melilotus officinalis (homoeothetic compound raceme)

Heterothetic compound raceme

Hebe albicans (heterothetic compound raceme)

Compound spike

Lolium temulentum (compound spike)

Compound capitulum

Echinops ritro (compound capitulum)

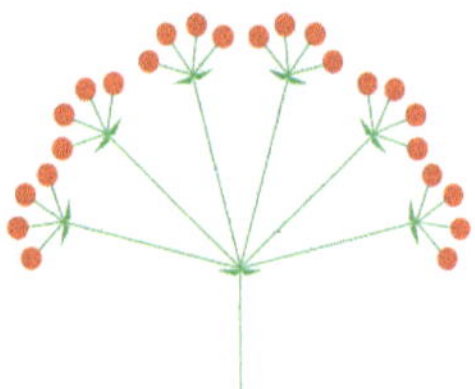

Compound (double) umbel

Laserpicium latifolium (double umbel)

Compound (triple) umbel

The most common kind of definite compound inflorescence is the **panicle** (of Webeling, or ‹panicle-like cyme'). A panicle is a definite inflorescence that is increasingly more strongly and irregularly branched from the top to the bottom and where each branching has a terminal flower.

The so-called cymose **corymb** is similar to a racemose corymb but has a panicle-like structure. Another type of panicle is the **anthela**. An anthela is a cymose corymb with the lateral flowers higher than the central ones.

Panicle

Vitis vinifera *(panicle)*

Cymose corymb

Sambucus nigra (cymose corymb)

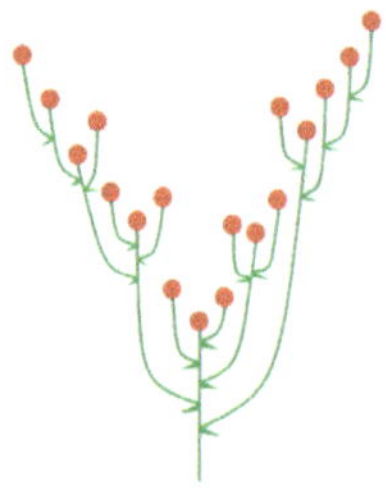

Anthela

Juncus inflexus (anthela)

A raceme in which the single flowers are replaced by cymes is called a (indefinite) **thyrse**. The secondary cymes can of course be of any of the different types of dichasia and monochasia. A botryoid in which the single flowers are replaced by cymes is a **definite thyrse** or **thyrsoid**. Thyrses are often confusingly called **panicles**.

Thyrse

Aesculus hippocastanum

Thyrsoid

Syringa vulgaris

Other combinations are, of course, possible. For example, heads or umbels may be arranged in a corymb or a panicle.

Achillea sp. (heads in a corymb)

Hedera helix (umbels in a panicle)

Other

The family Asteraceae is characterised by a highly specialised head technically called a calathid (but usually referred to as 'capitulum' or ‹head'). The family Poaceae has a peculiar inflorescence of small spikes (spikelets) organised in panicles or spikes that are usually simply and improperly referred to as spike and panicle. The genus *Ficus* (Moraceae) has an inflorescence called syconium and the genus *Euphorbia* has cyathia (sing. *cyathium*), usually organised in umbels. For detailed descriptions, see the respective articles.

Matricaria chamomilla (calathid)

Triticum aestivum (compound spikes, "spikes")

Oryza sativa (spikes in a panicle, “panicle”)

Ficus carica (syconium)

Euphorbia tridentata (cyathium)

Euphorbia cyparissias (cyathia in an umbel)

(*Coleus*-false spike)

FLOWER,

A **flower**, sometimes known as a bloom or blossom, is the reproductive structure found in plants that are floral (plants of the division Magnoliophyta, also called angiosperms). The biological function of a flower is to effect reproduction, usually by providing a mechanism for the union of sperm with eggs. Flowers may facilitate outcrossing (fusion of sperm and eggs from different individuals in a population) or allow selfing (fusion of sperm and egg from the same flower). Some flowers produce diaspores without fertilization (parthenocarpy). Flowers contain sporangia and are the site where gametophytes develop. Many flowers have evolved to be attractive to animals, so as to cause them to be vectors for the transfer of pollen. After fertilization, the ovary of the flower develops into fruit containing seeds.

In addition to facilitating the reproduction of flowering plants, flowers have long been admired and used by humans to beautify their environment, and also as objects of romance, ritual, religion, medicine and as a source of food.

Morphology

Floral Parts

The essential parts of a flower can be considered in two parts: the vegetative part, consisting of petals and associated structures in the perianth, and the reproductive or sexual parts. A stereotypical flower consists of four kinds of structures attached to the tip of a short stalk. Each of these kinds of parts is arranged in a whorl on the receptacle. The four main whorls (starting from the base of the flower or lowest node and working upwards) are as follows:

Main parts of a mature flower (Ranunculus glaberrimus)

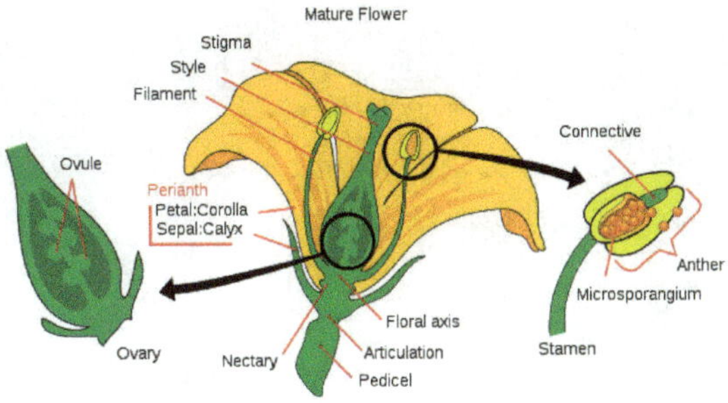

Diagram of flower parts

Perianth

Collectively the calyx and corolla form the perianth (see diagram).

- *Calyx*: the outermost whorl consisting of units called *sepals*; these are typically green and enclose the rest of the flower in the bud stage, however, they can be absent or prominent and petal-like in some species.
- *Corolla*: the next whorl toward the apex, composed of units called *petals*, which are typically thin, soft and colored to attract animals that help the process of pollination.

Reproductive

- *Androecium* (from Greek *andros oikia*: man's house): the next whorl (sometimes multiplied into several whorls), consisting of units called stamens. Stamens consist of two parts: a stalk called a filament, topped by an anther where pollen is produced by meiosis and eventually dispersed.
- *Gynoecium* (from Greek *gynaikos oikia*: woman's house): the innermost whorl of a flower, consisting of one or more units called carpels. The carpel or multiple fused carpels form a hollow structure called an ovary, which produces ovules internally. Ovules are megasporangia and they in turn produce megaspores by meiosis which develop into female gametophytes. These give rise to egg cells. The gynoecium of a flower is also described using an alternative terminology wherein the structure one sees in the innermost whorl (consisting of an ovary, style and stigma) is called a pistil. A pistil may consist of a single carpel or a number of carpels fused together. The sticky tip of the pistil, the stigma, is the receptor of pollen. The supportive stalk, the style, becomes the pathway for pollen tubes to grow from pollen grains adhering to the stigma. The relationship to the gynoecium on the receptacle is described as **hypogynous** (beneath a superior ovary), **perigynous** (surrounding a superior ovary), or **epigynous** (above inferior ovary).

Reproductive parts of Easter Lily (*Lilium longiflorum*). 1. Stigma, 2. Style, 3. Stamens, 4. Filament, 5. Petal

Structure

Although the arrangement described above is considered "typical", plant species show a wide variation in floral structure. These modifications have significance in the evolution of flowering plants and are used extensively by botanists to establish relationships among plant species.

The four main parts of a flower are generally defined by their positions on the receptacle and not by their function. Many flowers lack some parts or parts may be modified into other functions and/or look like what is typically another part. In some families, like Ranunculaceae, the petals are greatly reduced and in many species the sepals are colorful and petal-like. Other flowers have modified stamens that are petal-like; the double flowers of Peonies and Roses are mostly petaloid stamens.Flowers show great variation and plant scientists describe this variation in a systematic way to identify and distinguish species.

Specific terminology is used to describe flowers and their parts. Many flower parts are fused together; fused parts originating from the same whorl are **connate**, while fused parts originating from different whorls are **adnate**; parts that are not fused are **free**. When petals are fused into a tube or ring that falls away as a single unit, they are sympetalous (also called **gamopetalous**). Connate petals may have distinctive regions: the cylindrical base is the tube, the expanding region is the throat and the flaring outer region is the limb. A sympetalous flower, with bilateral symmetry with an upper and lower lip, is **bilabiate**. Flowers with connate petals or sepals may have various shaped corolla or calyx, including campanulate, funnelform, tubular, urceolate, salverform or rotate.

Referring to "fusion," as it is commonly done, appears questionable because at least some of the processes involved may be non-fusion processes. For example, the addition of intercalary growth at or below the base of the primordia of floral appendages such as sepals, petals, stamens and carpels may lead to a common base that is not the result of fusion.

Left: A normal zygomorphic Streptocarpus flower. Right: An aberrant peloric Streptocarpus flower. Both of these flowers appeared on the Streptocarpus hybrid 'Anderson's Crows' Wings'.

Many flowers have a symmetry. When the perianth is bisected through the central axis from any point, symmetrical halves are produced, forming a radial symmetry. These flowers are also known to be actinomorphic or regular, e.g. rose or trillium. When flowers are bisected and produce only one line that produces symmetrical halves the flower is said to be irregular or zygomorphic, e.g. snapdragon or most orchids.

Flowers may be directly attached to the plant at their base (sessile—the supporting stalk or stem is highly reduced or absent). The stem or stalk subtending a flower is called a peduncle. If a peduncle supports more than one flower, the stems connecting each flower to the main axis are called pedicels. The apex of a flowering stem forms a terminal swelling which is called the *torus* or receptacle.

Inflorescence

In those species that have more than one flower on an axis, the collective cluster of flowers is termed an *inflorescence*. Some inflorescences are composed of many small flowers arranged in a formation that resembles a single flower. The common example of this is most members of the very large composite (Asteraceae) group. A single daisy or sunflower, for example, is not a flower but a flower *head*—an inflorescence composed of numerous flowers (or florets). An inflorescence may include specialized stems and modified leaves known as bracts.

The familiar calla lily is not a single flower. It is actually an inflorescence of tiny flowers pressed together on a central stalk that is surrounded by a large petal-like bract.

Floral Diagrams and Floral Formulae

A *floral formula* is a way to represent the structure of a flower using specific letters, numbers and symbols, presenting substantial information about the flower in a compact form. It can represent a taxon, usually giving ranges of the numbers of different organs, or particular species. Floral formulae have been developed in the early 19th century and their use has declined since. Prenner *et al.* (2010) devised an extension of the existing model to broaden the descriptive capability of the formula. The format of floral formulae differs in different parts of the world, yet they convey the same information.

The structure of a flower can also be expressed by the means of *floral diagrams*. The use of schematic diagrams can replace long descriptions or complicated drawings as a tool for understanding both floral structure and evolution. Such diagrams may show important features of flowers, including the relative positions of the various organs, including the presence of fusion and symmetry, as well as structural details.

Development

A flower develops on a modified shoot or *axis* from a determinate apical meristem (*determinate* meaning the axis grows to a set size). It has compressed internodes, bearing structures that in classical plant morphology are interpreted as highly modified leaves. Detailed developmental studies, however, have shown that stamens are often initiated more or less like modified stems (caulomes) that in some cases may even resemble branchlets. Taking into account the whole diversity in the development of the androecium of flowering plants, we find a continuum between modified leaves (phyllomes), modified stems (caulomes), and modified branchlets (shoots).

Flowering Transition

The transition to flowering is one of the major phase changes that a plant makes during its life cycle. The transition must take place at a time that is favorable for fertilization and the formation of seeds, hence ensuring maximal reproductive success. To meet these needs a plant is able to interpret important endogenous and environmental cues such as changes in levels of plant hormones and seasonable temperature and photoperiod changes. Many perennial and most biennial plants require vernalization to flower. The molecular interpretation of these signals is through the transmission of a complex signal known as florigen, which involves a variety of genes, including CONSTANS, FLOWERING LOCUS C and FLOWERING LOCUS T. Florigen is produced in the leaves in reproductively favorable conditions and acts in buds and growing tips to induce a number of different physiological and morphological changes.

The first step of the transition is the transformation of the vegetative stem primordia into floral primordia. This occurs as biochemical changes take place to change cellular differentiation of leaf, bud and stem tissues into tissue that will grow into the reproductive organs. Growth of the central part of the stem tip stops or flattens out and the sides develop protuberances in a whorled or spiral fashion around the outside of the stem end. These protuberances develop into the sepals, petals, stamens, and carpels. Once this process begins, in most plants, it cannot be reversed and the stems develop flowers, even if the initial start of the flower formation event was dependent of some environmental cue. Once the process begins, even if that cue is removed the stem will continue to develop a flower.

Organ Development

The molecular control of floral organ identity determination appears to be fairly well understood in some species. In a simple model, three gene activities interact in a combinatorial manner to determine the developmental identities of the organ primordia within the floral meristem. These gene functions are called A, B and C-gene functions. In the first floral whorl only A-genes are expressed, leading to the formation of sepals. In the second whorl both A- and B-genes are expressed, leading to the formation of petals. In the third whorl, B and C genes interact to form stamens and in the center of the flower C-genes alone give rise to carpels. The model is based upon studies of homeotic mutants in *Arabidopsis thaliana* and snapdragon, *Antirrhinum majus*. For example, when there is a loss of B-gene function, mutant flowers are produced with sepals in the first whorl as usual, but also in the second whorl instead of the normal petal formation. In the third whorl the lack of B function but presence of C-function mimics the fourth whorl, leading to the formation of carpels also in the third whorl. See also The ABC Model of Flower Development.

Most genes central in this model belong to the MADS-box genes and are transcription factors that regulate the expression of the genes specific for each floral organ.

Floral Function

The principal purpose of a flower is the reproduction of the individual and the species. All flowering plants are *heterosporous*, producing two types of spores. Microspores are produced by meiosis inside anthers while megaspores are produced inside ovules, inside an ovary. In fact, anthers typically consist of four microsporangia and an ovule is an integumented megasporangium. Both types of spores develop into gametophytes inside sporangia. As with all heterosporous plants, the gametophytes also develop inside the spores (are endosporic).

In the majority of species, individual flowers have both functional carpels and stamens. Botanists describe these flowers as being *perfect* or *bisexual* and the species as *hermaphroditic*. Some flowers lack one or the other reproductive organ and called *imperfect* or *unisexual*. If unisex flowers are found on the same individual plant but in different locations, the species is said to be *monoecious*. If each type of unisex flower is found only on separate individuals, the plant is *dioecious*.

An example of a "perfect flower", this *Crateva religiosa* flower has both stamens (outer ring) and a pistil (center).

Flower Specialization and Pollination

Flowering plants usually face selective pressure to optimize the transfer of their pollen, and this is typically reflected in the morphology of the flowers and the behaviour of the plants. Pollen may be transferred between plants via a number of 'vectors'. Some plants make use of abiotic vectors — namely wind (anemophily) or, much less commonly, water (hydrophily). Others use biotic vectors including insects (entomophily), birds (ornithophily), bats (chiropterophily) or other animals. Some plants make use of multiple vectors, but many are highly specialised.

Cleistogamous flowers are self-pollinated, after which they may or may not open. Many Viola and some Salvia species are known to have these types of flowers.

The flowers of plants that make use of biotic pollen vectors commonly have glands called nectaries that act as an incentive for animals to visit the flower. Some flowers have patterns, called nectar guides, that show pollinators where to look for nectar. Flowers also attract pollinators by scent and color. Still other flowers use mimicry to attract pollinators. Some species of orchids, for example, produce flowers resembling female bees in color, shape, and scent. Flowers are also specialized in shape and have an arrangement of the stamens that ensures that pollen grains are transferred to the bodies of the pollinator when it lands in search of its attractant (such as nectar, pollen, or a mate). In pursuing this attractant from many flowers of the same species, the pollinator transfers pollen to the stigmas—arranged with equally pointed precision—of all of the flowers it visits.

Anemophilous flowers use the wind to move pollen from one flower to the next. Examples include grasses, birch trees, ragweed and maples. They have no need to attract pollinators and therefore tend not to be "showy" flowers. Male and female reproductive organs are generally found in separate flowers, the male flowers having a number of long filaments terminating in exposed stamens, and the female flowers having long, feather-like stigmas. Whereas the pollen of animal-pollinated flowers tends to be large-grained, sticky, and rich in protein (another "reward" for pollinators), anemophilous flower pollen is usually small-grained, very light, and of little nutritional value to animals.

Pollination

Grains of pollen sticking to this bee will be transferred to the next flower it visits

The primary purpose of a flower is reproduction. Since the flowers are the reproductive organs of plant, they mediate the joining of the sperm, contained within pollen, to the ovules — contained in the ovary. Pollination is the movement of pollen from the anthers to the stigma. The joining of the sperm to the ovules is called fertilization. Normally pollen is moved from one plant to another, but many plants are able to self pollinate. The fertilized ovules produce seeds that are the next generation. Sexual reproduction produces genetically unique offspring, allowing for adaptation. Flowers have specific designs which encourages the transfer of pollen from one plant to another of the same species. Many plants are dependent upon external factors for pollination, including: wind and animals, and especially insects. Even large animals such as birds, bats, and pygmy possums can be employed. The period of time during which this process can take place (the flower is fully expanded and functional) is called *anthesis*. The study of pollination by insects is called *anthecology*.

Pollen

The types of pollen that most commonly cause allergic reactions are produced by the plain-looking plants (trees, grasses, and weeds) that do not have showy flowers. These plants make small, light, dry pollen grains that are custom-made for wind transport.

The type of allergens in the pollen is the main factor that determines whether the pollen is likely to cause hay fever. For example, pine tree pollen is produced in large amounts by a common tree, which would make it a good candidate for causing allergy. It is, however, a relatively rare cause of allergy because the types of allergens in pine pollen appear to make it less allergenic.

Among North American plants, weeds are the most prolific producers of allergenic pollen. Ragweed is the major culprit, but other important sources are sagebrush, redroot pigweed, lamb's quarters, Russian thistle (tumbleweed), and English plantain.

There is much confusion about the role of flowers in allergies. For example, the showy and entomophilous goldenrod (*Solidago*) is frequently blamed for respiratory allergies, of which it is innocent, since its pollen cannot be airborne. Instead the allergen is usually the pollen of the contemporary bloom of anemophilous ragweed (*Ambrosia*), which can drift for many kilometers.

Scientists have collected samples of ragweed pollen 400 miles out at sea and 2 miles high in the air. A single ragweed plant can generate a million grains of pollen per day.

It is common to hear people say they are allergic to colorful or scented flowers like roses. In fact, only florists, gardeners, and others who have prolonged, close contact with flowers are likely to be sensitive to pollen from these plants. Most people have little contact with the large, heavy, waxy pollen grains of such flowering plants because this type of pollen is not carried by wind but by insects such as butterflies and bees.

Attraction Methods

Plants cannot move from one location to another, thus many flowers have evolved to attract animals to transfer pollen between individuals in dispersed populations. Flowers that are insect-pollinated are called *entomophilous*; literally "insect-loving" in Greek. They can be highly modified along with the pollinating insects by co-evolution. Flowers commonly have glands called *nectaries* on various parts that attract animals looking for nutritious nectar. Birds and bees have color vision, enabling them to seek out "colorful" flowers.

A Bee orchid has evolved over many generations to better mimic a female bee to attract male bees as pollinators.

Some flowers have patterns, called nectar guides, that show pollinators where to look for nectar; they may be visible only under ultraviolet light, which is visible to bees and some other insects. Flowers also attract pollinators by scent and some of those scents are pleasant to our sense of smell. Not all flower scents are appealing to humans; a number of flowers are pollinated by insects that are attracted to rotten flesh and have flowers that smell like dead animals, often called Carrion flowers, including *Rafflesia*, the titan arum, and the North American pawpaw (*Asimina triloba*). Flowers pollinated by night visitors, including bats and moths, are likely to concentrate on scent to attract pollinators and most such flowers are white.

Other flowers use mimicry to attract pollinators. Some species of orchids, for example, produce flowers resembling female bees in color, shape, and scent. Male bees move from one such flower to another in search of a mate.

Pollination Mechanism

The pollination mechanism employed by a plant depends on what method of pollination is utilized.

Most flowers can be divided between two broad groups of pollination methods:

Entomophilous: flowers attract and use insects, bats, birds or other animals to transfer pollen from one flower to the next. Often they are specialized in shape and have an arrangement of the stamens that ensures that pollen grains are transferred to the bodies of the pollinator when it lands in search of its attractant (such as nectar, pollen, or a mate). In pursuing this attractant from many flowers of the same species, the pollinator transfers pollen to the stigmas—arranged with equally pointed precision—of all of the flowers it visits. Many flowers rely on simple proximity between flower parts to ensure pollination. Others, such as the *Sarracenia* or lady-slipper orchids, have elaborate designs to ensure pollination while preventing self-pollination.

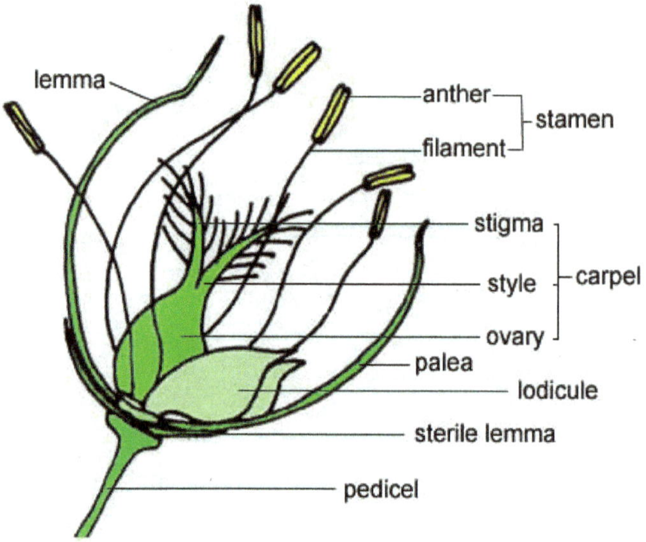

Grass flower with vestigial perianth or lodicules

Anemophilous: flowers use the wind to move pollen from one flower to the next, examples include the grasses, Birch trees, Ragweed and Maples. They have no need to attract pollinators and therefore tend not to grow large blossoms. Whereas the pollen of entomophilous flowers tends to be large-grained, sticky, and rich in protein (another "reward" for pollinators), anemophilous flower pollen is usually small-grained, very light, and of little nutritional value to insects, though it may still be gathered in times of dearth. Honeybees and bumblebees actively gather anemophilous corn (maize) pollen, though it is of little value to them.

Some flowers are self-pollinated and use flowers that never open or are self-pollinated before the flowers open, these flowers are called cleistogamous. Many Viola species and some Salvia have these types of flowers.

Flower-Pollinator Relationships

Many flowers have close relationships with one or a few specific pollinating organisms. Many flowers, for example, attract only one specific species of insect, and therefore rely on that insect for successful reproduction. This close relationship is often given as an example of coevolution, as the flower and pollinator are thought to have developed together over a long period of time to match each other's needs.

This close relationship compounds the negative effects of extinction. The extinction of either member in such a relationship would mean almost certain extinction of the other member as well. Some endangered plant species are so because of shrinking pollinator populations.

Fertilization and Dispersal

Some flowers with both stamens and a pistil are capable of self-fertilization, which does increase the chance of producing seeds but limits genetic variation. The extreme case of self-fertilization occurs in flowers that always self-fertilize, such as many dandelions. Conversely, many species of plants have ways of preventing self-fertilization. Unisexual male and female flowers on the same plant may not appear or mature at the same time, or pollen from the same plant may be incapable of fertilizing its ovules. The latter flower types, which have chemical barriers to their own pollen, are referred to as self-sterile or self-incompatible.

EVOLUTION

While land plants have existed for about 425 million years, the first ones reproduced by a simple adaptation of their aquatic counterparts: spores. In the sea, plants—and some animals—can simply scatter out genctic clones of themselves to float away and grow elsewhere. This is how early plants reproduced. But plants soon evolved methods of protecting these copies to deal with drying out and other abuse which is even more likely on land than in the sea. The protection became the seed, though it had not yet evolved the flower. Early seed-bearing plants include the ginkgo and conifers. The earliest fossil of a flowering plant, *Archaefructus liaoningensis*, is dated about 125 million years old.

Archaefructus liaoningensis, one of the earliest known flowering plants

Several groups of extinct gymnosperms, particularly seed ferns, have been proposed as the ancestors of flowering plants but there is no continuous fossil evidence showing exactly how flowers evolved. The apparently sudden appearance of relatively modern flowers in the fossil record posed such a problem for the theory of evolution that it was called an "abominable mystery" by Charles Darwin. Recently discovered angiosperm fossils such as *Archaefructus*, along with further discoveries of fossil gymnosperms, suggest how angiosperm characteristics may have been acquired in a series of steps.

Recent DNA analysis (molecular systematics) shows that *Amborella trichopoda*, found on the Pacific island of New Caledonia, is the sister group to the rest of the flowering plants, and morphological studiessuggest that it has features which may have been characteristic of the earliest flowering plants.

The general assumption is that the function of flowers, from the start, was to involve animals in the reproduction process. Pollen can be scattered without bright colors and obvious shapes, which would therefore be a liability, using the plant's resources, unless they provide some other benefit. One proposed reason for the sudden, fully developed appearance of flowers is that they evolved in an isolated setting like an island, or chain of islands, where the plants bearing them were able to develop a highly specialized relationship with some specific animal (a wasp, for example), the way many island species develop today. This symbiotic relationship, with a hypothetical wasp bearing pollen from one plant to another much the way fig wasps do today, could have eventually resulted in both the plant(s) and their partners developing a high degree of specialization. Island genetics is believed to be a common source of speciation, especially when it comes to radical adaptations which seem to have required inferior transitional forms. Note that the wasp example is not incidental; bees, apparently evolved specifically for symbiotic plant relationships, are descended from wasps.

Likewise, most fruit used in plant reproduction comes from the enlargement of parts of the flower. This fruit is frequently a tool which depends upon animals wishing to eat it, and thus scattering the seeds it contains.

While many such symbiotic relationships remain too fragile to survive competition with mainland organisms, flowers proved to be an unusually effective means of production, spreading (whatever their actual origin) to become the dominant form of land plant life.

Amborella trichopoda, the sister group to the rest of the flowering plants

While there is only hard proof of such flowers existing about 130 million years ago, there is some circumstantial evidence that they did exist up to 250 million years ago. A chemical used by plants to defend their flowers, oleanane, has been detected in fossil plants that old, including gigantopterids, which evolved at that time and bear many of the traits of modern, flowering plants, though they are not known to be flowering plants themselves, because only their stems and prickles have been found preserved in detail; one of the earliest examples of petrification.

The similarity in leaf and stem structure can be very important, because flowers are genetically just an adaptation of normal leaf and stem components on plants, a combination of genes normally responsible for forming new shoots. The most primitive flowers are thought to have had a variable number of flower parts, often separate from (but in contact with) each other. The flowers would have tended to grow in a spiral pattern, to be bisexual (in plants, this means both male and female parts on the same flower), and to be dominated by the ovary (female part). As flowers grew more advanced, some variations developed parts fused together, with a much more specific number and design, and with either specific sexes per flower or plant, or at least "ovary inferior".

Flower evolution continues to the present day; modern flowers have been so profoundly influenced by humans that many of them cannot be pollinated in nature. Many modern, domesticated flowers used to be simple weeds, which only sprouted when the ground was disturbed. Some of them tended to grow with human crops, and the prettiest did not get plucked because of their beauty, developing a dependence upon and special adaptation to human affection.

Color

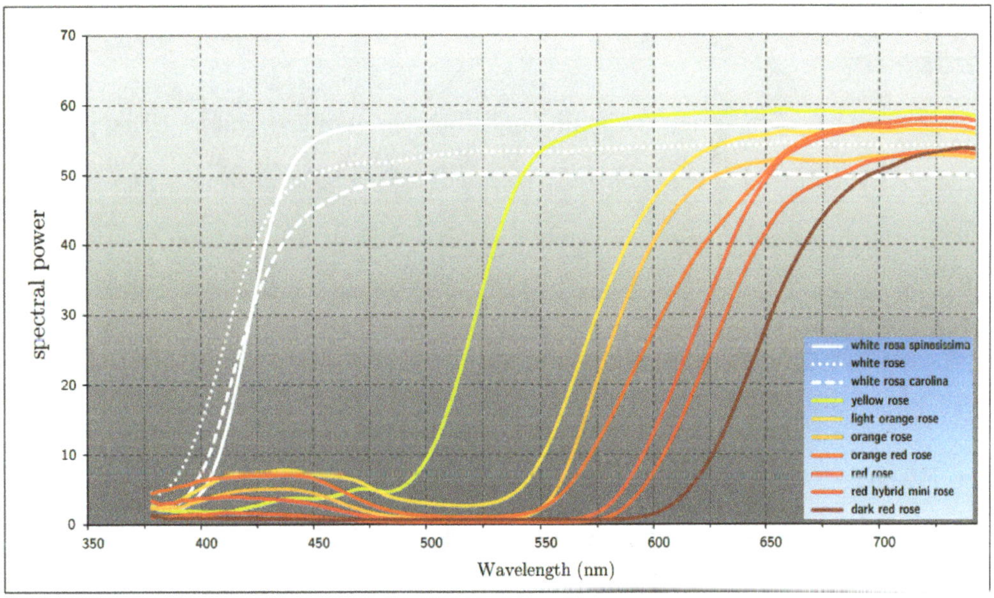

Spectrum of the flowers of the Rose family of plants. A red rose absorbs about 99.7% of light across a broad area below the red wavelengths of the spectrum, leading to an exceptionally *pure* red. A yellow rose will reflect about 5% of blue light, producing an unsaturated yellow (a yellow with a degree of white in it).

Many flowering plants reflect as much light as possible within the range of visible wavelengths of the pollinator the plant intends to attract. Flowers that reflect the full range of visible light are generally perceived as *white* by a human observer. An important feature of white flowers is that they reflect equally across the visible spectrum. While many flowering plants use white to attract pollinators, the use of color is also widespread (even within the same species). Color allows a flowering plant to be more specific about the pollinator it seeks to attract. The color model used by human color reproduction technology (CMYK) relies on the modulation of pigments that divide the spectrum into broad areas of absorption. Flowering plants by contrast are able to shift the transition point wavelength between absorption and reflection. If it is assumed that the visual systems of most pollinators view the visible spectrum as circular then it may be said that flowering plants produce color by absorbing the light in one region of the spectrum and reflecting the light in the other region. With CMYK, color is produced as a function of the amplitude of the broad regions of absorption. Flowering plants by contrast produce color by modifying the frequency (or rather wavelength) of the light reflected. Most flowers absorb light in the blue to yellow region of the spectrum and reflect light from the green to red region of the spectrum. For many species of flowering plant, it is the transition point that characterizes the color that they produce. Color may be modulated by shifting the transition point between absorption and reflection and in this way a flowering plant may specify which pollinator it seeks to attract.

Some flowering plants also have a limited ability to modulate areas of absorption. This is typically not as precise as control over wavelength. Humans observers will perceive this as degrees of saturation (the amount of *white* in the color).

Symbolism

Many flowers have important symbolic meanings in Western culture. The practice of assigning meanings to flowers is known as floriography. Some of the more common examples include:

- Red roses are given as a symbol of love, beauty, and passion.
- Poppies are a symbol of consolation in time of death. In the United Kingdom, New Zealand, Australia and Canada, red poppies are worn to commemorate soldiers who have died in times of war.
- Irises/Lily are used in burials as a symbol referring to "resurrection/life". It is also associated with stars (sun) and its petals blooming/shining.
- Daisies are a symbol of innocence.

Red Rose

Lilies are often used to denote life or resurrection

Flowers are common subjects of still life paintings, such as this one by Ambrosius Bosschaert the Elder

Flowers within art are also representative of the female genitalia, as seen in the works of artists such as Georgia O'Keeffe, Imogen Cunningham, Veronica Ruiz de Velasco, and Judy Chicago, and in fact in Asian and western classical art. Many cultures around the world have a marked tendency to associate flowers with femininity.

The great variety of delicate and beautiful flowers has inspired the works of numerous poets, especially from the 18th-19th century Romantic era. Famous examples include

William Wordsworth's *I Wandered Lonely as a Cloud* and William Blake's *Ah! Sun-Flower*.

Because of their varied and colorful appearance, flowers have long been a favorite subject of visual artists as well. Some of the most celebrated paintings from well-known painters are of flowers, such as Van Gogh's sunflowers series or Monet's water lilies. Flowers are also dried, freeze dried and pressed in order to create permanent, three-dimensional pieces of flower art.

Their symbolism in dreams has also been discussed, with possible interpretations including "blossoming potential".

The Roman goddess of flowers, gardens, and the season of Spring is Flora. The Greek goddess of spring, flowers and nature is Chloris.

In Hindu mythology, flowers have a significant status. Vishnu, one of the three major gods in the Hindu system, is often depicted standing straight on a lotus flower.Apart from the association with Vishnu, the Hindu tradition also considers the lotus to have spiritual significance. For example, it figures in the Hindu stories of creation.

Usage

In modern times people have sought ways to cultivate, buy, wear, or otherwise be around flowers and blooming plants, partly because of their agreeable appearance and smell. Around the world, people use flowers for a wide range of events and functions that, cumulatively, encompass one's lifetime:

- For new births or christenings
- As a corsage or boutonniere worn at social functions or for holidays
- As tokens of love or esteem
- For wedding flowers for the bridal party, and for decorations for the hall
- As brightening decorations within the home
- As a gift of remembrance for *bon voyage* parties, welcome-home parties, and "thinking of you" gifts
- For funeral flowers and expressions of sympathy for the grieving
- For worshiping goddesses. In Hindu culture adherents commonly bring flowers as a gift to temples

People therefore grow flowers around their homes, dedicate entire parts of their living space to flower gardens, pick wildflowers, or buy flowers from florists who depend on an entire network of commercial growers and shippers to support their trade.

Flower market - Detroit's Eastern Market

A woman spreading flowers over a lingam in a temple in Varanasi

Flowers provide less food than other major plants parts (seeds, fruits, roots, stems and leaves) but they provide several important foods and spices. Flower vegetables include broccoli, cauliflower and artichoke. The most expensive spice, saffron, consists of dried stigmas of a crocus. Other flower spices are cloves and capers. Hops flowers are used to flavor beer. Marigold flowers are fed to chickens to give their egg yolks a golden yellow color, which consumers find more desirable; dried and ground marigold flowers are also used as a spice and colouring agent in Georgian cuisine. Flowers of the dandelion and elder are often made into wine. Bee pollen, pollen collected from bees, is considered a health food by some people. Honey consists of bee-processed flower nectar and is often named for the type of flower, e.g. orange blossom honey, clover honey and tupelo honey.

Hundreds of fresh flowers are edible but few are widely marketed as food. They are often used to add color and flavor to salads. Squash flowers are dipped in breadcrumbs and fried. Edible flowers include nasturtium, chrysanthemum, carnation, cattail, honeysuckle, chicory, cornflower, canna, and sunflower. Some edible flowers are sometimes candied such as daisy, rose, and violet (one may also come across a candied pansy).

Flowers can also be made into herbal teas. Dried flowers such as chrysanthemum, rose, jasmine, camomile are infused into tea both for their fragrance and medical properties. Sometimes, they are also mixed with tea leaves for the added fragrance.

Flowers have been used since as far back as 50,000 years in funeral rituals. Many cultures do draw a connection between flowers and life and death, and because of their seasonal return flowers also suggest rebirth, which may explain why many people place flowers upon graves. In ancient times the Greeks would place a crown of flowers on the head of the deceased as well as cover tombs with wreaths and flower petals. Rich and powerful women in ancient Egypt would wear floral headdresses and necklaces upon their death as representations of renewal and a joyful afterlife, and the Mexicans to this day use flowers prominently in their Day of the Dead celebrations in the same way that their Aztec ancestors did.

Eight Flowers, a painting by artist Qian Xuan, 13th century, Palace Museum, Beijing.

PERIANTH

The **perianth** (sometimes called perigonium or perigon) is the non-reproductive part of the flower, and structure that forms an envelope surrounding the sexual organs, consisting of the calyx (sepals) and the corolla (petals). The term perianth is derived from the Greek περί, *peri*, meaning around, and άνθος, *anthos*, meaning flower, while perigonium is derived from *gonos*, meaning seed, i.e. sexual organs. In the mosses and liverworts (Marchantiophyta), the perianth is the sterile tubelike tissue that surrounds the female reproductive structure (or developing sporophyte).

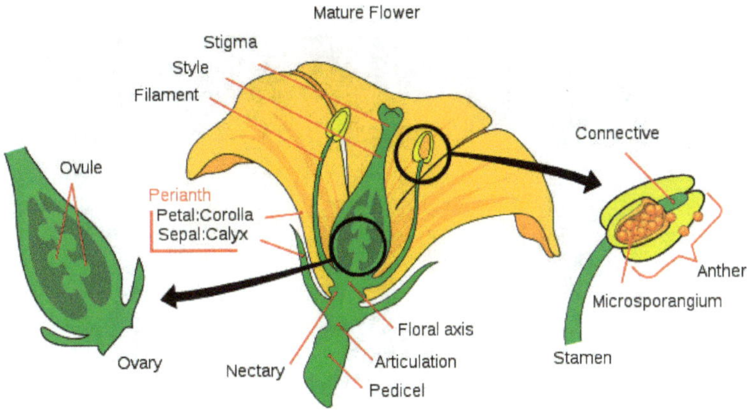

Diagram showing the parts of a mature flower. In this example the perianth is separated into a calyx (sepals) and corolla (petals)

Flowering Plants

In flowering plants, the perianth may be described as being either **dichlamydeous/ heterochlamydeous** in which the calyx and corolla are clearly separate, or **homochlamydeous**, in which they are indistinguishable (and the sepals and petals are collectively referred to as **tepals**). When the perianth is in two whorls, it is described as **biseriate**. While the calyx may be green, it may also be brightly coloured, and is then described as **petaloid**. When the undifferentiated tepals resemble petals, they are also referred to as "petaloid", as in petaloid monocots, orders of monocots with brightly coloured tepals. Since they include Liliales, an alternative name is lilioid monocots. The corolla and petals have a role in attracting pollinators, but this may be augmented by more specialised structures like the **corona** (see below).

When the corolla consists of separate tepals the term **apotepalous** is used, or **syntepalous** if the tepals are fused to one another. The petals may be united to form a tubular corolla (**gamopetalous** or **sympetalous**). If either the petals or sepals are entirely absent, the perianth can be described as being **monochlamydeous**.

Corona

Flower of *Narcissus* showing an outer white corolla with a central yellow corona (paraperigonium)

Flower of *Passiflora incarnata* showing corona of fine appendages between petals and stamens

An additional structure in some plants (e.g. *Narcissus*, *Passiflora* (passion flower), some *Hippeastrum*, Liliaceae) is the **corona** (paraperigonium, paraperigon, or paracorolla), a ring or set of appendages of adaxial tissue arising from the corolla or the outer edge of the stamens. It is often positioned where the corolla lobes arise from the corolla tube.

The pappus of Asteraceae, considered to be a modified calyx, is also called a corona if it is shaped like a crown.

Types of Perianth

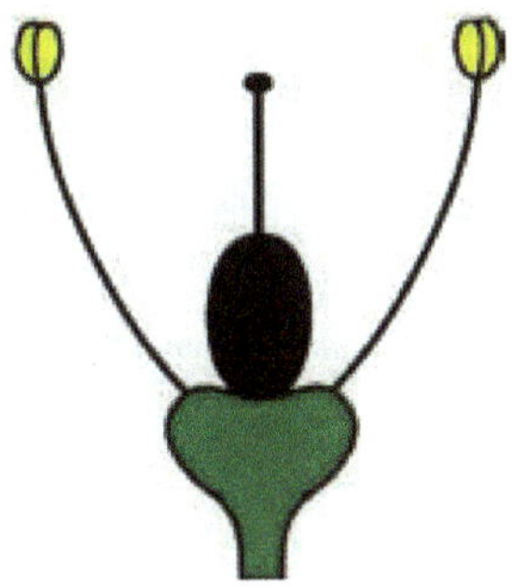

Achlamydeous floral meristem without a corolla or calyx

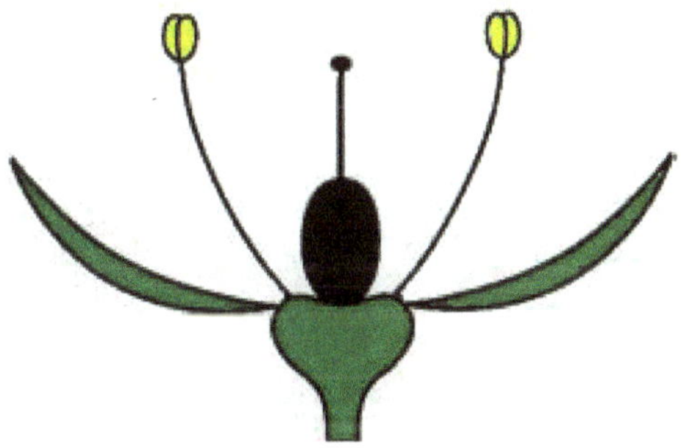

Monochlamydeous perianth with non-petaloid calyx only

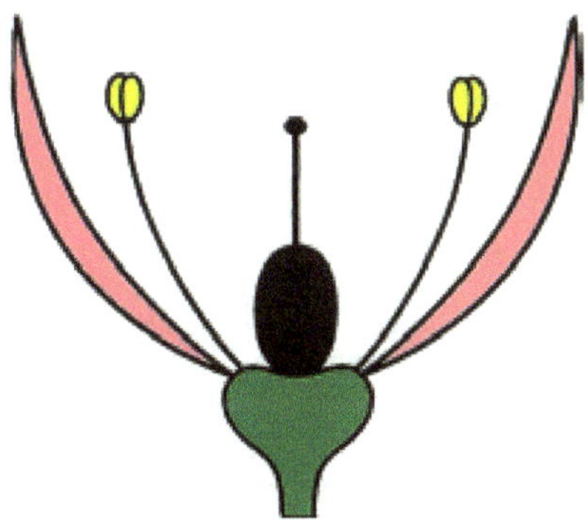

Monochlamydeous perianth with corolla only or homochlamydeous perigonium with tepals

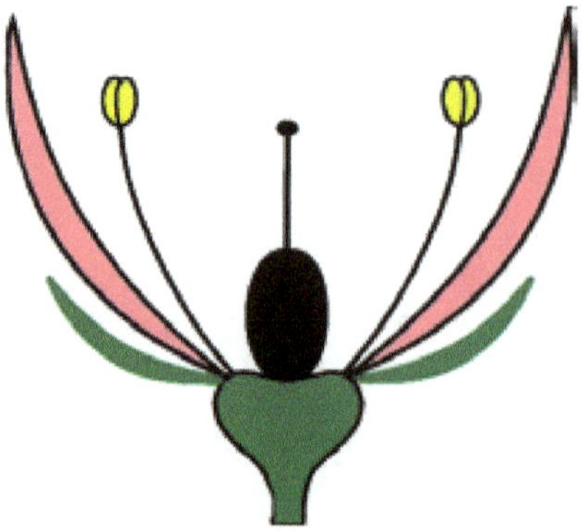

Dichlamydeous/heterochlamydeous perianth with separate whorls

Both sepals and petals may have stomata and veins, even if vestigial. In some taxa, for instance some magnolias and water lilies the perianth is arranged in a spiral on nodes, rather than whorls. Flowers with spiral perianths tend to also be those with undifferentiated perianths.

Corona

Flower of *Narcissus* showing an outer white corolla with a central yellow corona (paraperigonium)

Flower of *Passiflora incarnata* showing corona of fine appendages between petals and stamens

An additional structure in some plants (e.g. *Narcissus*, *Passiflora* (passion flower), some *Hippeastrum*, Liliaceae) is the **corona** (paraperigonium, paraperigon, or paracorolla), a ring or set of appendages of adaxial tissue arising from the corolla or the outer edge of the stamens. It is often positioned where the corolla lobes arise from the corolla tube.

The pappus of Asteraceae, considered to be a modified calyx, is also called a corona if it is shaped like a crown.

SPOROPHYLL

A sporophyll is a leaf that bears sporangia. Both microphylls and megaphylls can be sporophylls. In heterosporous plants, sporophylls (whether they are microphylls or megaphylls) bear either megasporangia and thus are called megasporophylls, or microsporangia and are called microsporophylls. The overlap of the prefixes and roots makes these terms a particularly confusing subset of botanical nomenclature.

The sporophyll of a fern. It is a fertile leaf bearing reproductive structures.

Sporophylls vary greatly in appearance and structure, and may or may not look similar to sterile leaves. Plants that produce sporophylls include:

Alaria esculenta a brown alga shows sporophylls attached near the base of the alga.

Lycophytes, where sporophylls may be aggregated into strobili (*Selaginella* and some *Lycopodium* and related genera) or distributed singly among sterile leaves (*Huperzia*). Sporangia are borne in the axil or on the adaxial surface of the sporophyll. In heterosporous members, megasporophylls and microsporophylls may be intermixed or separated in a variety of patterns.

Ferns, which may produce sporophylls that are similar to sterile fronds or that appear very different from sterile fronds. These may be non-photosynthetic and lack typical pinnae (e.g. *Onoclea*)

Cycads produce strobili, both pollen-producing and seed-producing, that are composed of sporophylls.

Ginkgo produces microsporophylls aggregated into a pollen strobilus. Ovules are not born on sporophylls.

Conifers, like *Ginkgo* and cycads, produce microsporophylls, aggregated into pollen strobili. However, unlike these other groups, ovules are produced on cone scales, which are modified shoots rather than sporophylls.

Some plants do not produce sporophylls. Sporangia are produced directly on stems. *Psilotum* has been interpreted as producing sporangia (fused in a synangium) on the terminus of a stem. *Equisetum* always produce strobili, but the structures bearing sporangia (sporangiophores) have been interpreted as modified stems. The sporangia, despite being recurved are interpreted as terminal.

Gnetophytes produce both compound pollen and seed strobili.

ARCHEGONIUM

An **archegonium** (pl: **archegonia**), from the ancient Greek *ρχή* («beginning») and *γόνος* («offspring»), is a multicellular structure or organ of the gametophyte phase of certain plants, producing and containing the ovum or female gamete. The corresponding male organ is called the antheridium. The archegonium has a long neck canal or venter and a swollen base. Archegonia are typically located on the surface of the plant thallus, although in the hornworts they are embedded.

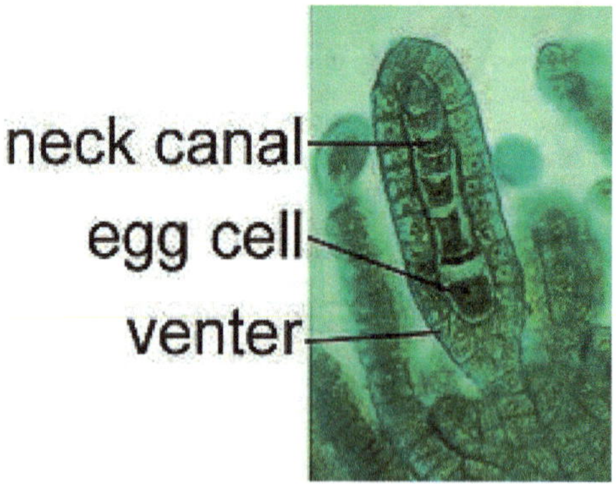

Diagram of archegonium anatomy

Bryophytes

In bryophytes and other cryptogams sperm reach the archegonium by swimming in water films, whereas in Pinophyta and Angiosperms the pollen are delivered by wind or animal vectors and the sperm are delivered by means of a pollen tube.

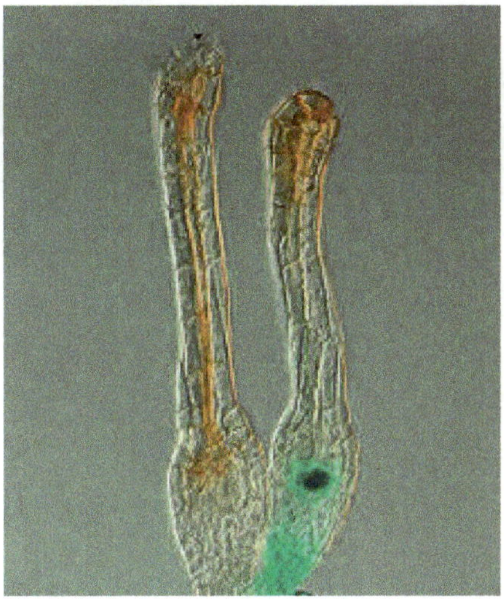

Gene expression pattern determined by histochemical GUS assays in *Physcomitrella patens*

In the moss *Physcomitrella patens*, archegonia are not embedded but are located on top of the leafy gametophore (s. Figure). The Polycomb protein FIE is expressed in the unfertilized egg cell (right) as the blue colour after GUS staining reveals. Soon after fertilisation the FIE gene is inactivated (the blue colour is no longer visible, left) in the young embryo.

Gymnosperms

They are much-reduced and embedded in the megagametophytes of gymnosperms. The term is not used for angiosperms or the gnetophytes *Gnetum* and *Welwitschia* because the megagametophyte is reduced to just a few cells, one of which differentiates into the egg cell. The function of surrounding the gamete is assumed in large part by diploid cells of the megasporangium (nucellus) inside the ovule. Gymnosperms have their archegonium formed after pollination inside female conifer cones (megastrobili).

STAMEN

From Wikipedia, the free encyclopedia

For the Bulgarian physician, see Stamen Grigorov. For the data visualization and cartography studio, see Stamen Design.

Androecium redirects here; for the structure in cryptogams, see Antheridium.

Stamens of a *Hippeastrum* with white filaments and prominent anthers carrying pollen

The **stamen** (plural *stamina* or *stamens*) is the pollen-producing reproductive organ of a flower. Collectively the stamens form the **androecium**.

Morphology and Terminology

A stamen typically consists of a stalk called the **filament** and an **anther** which contains *microsporangia*. Most commonly anthers are two-lobed and are attached to the filament either at the base or in the middle area of the anther. The sterile tissue between the lobes is called the **connective**. A pollen grain develops from a microspore in the microsporangium and contains the male gametophyte.

The stamens in a flower are collectively called the ***androecium***. The androecium in various species of plants forms a great variety of patterns, some of them highly complex. It surrounds the gynoecium and is surrounded by the perianth. A few members of the family Triuridaceae, particularly *Lacandonia schismatica*, are exceptional in that their gynoecia surround their androecia.

Hippeastrum flowers showing stamens above the style with its terminal stigma

Etymology

- ***Stamen*** is the Latin word meaning "thread" (originally thread of the warp, in weaving).

- **Filament** derives from classical Latin *filum*, meaning "thread"
- **Anther** derives from French *anthère*, from classical Latin *anthera*, meaning "medicine extracted from the flower" in turn from Ancient Greek νθηρά,feminine of νθηρός, "flowery", from νθος, "flower"
- **Androecium** derives from Ancient Greek νήρ meaning "man", and οκος meaning "house" or "chamber/room".

Variation in Morphology

Depending on the species of plant, some or all of the stamens in a flower may be attached to the petals or to the floral axis. They also may be free-standing or fused to one another in many different ways, including fusion of some but not all stamens. The filaments may be fused and the anthers free, or the filaments free and the anthers fused. Rather than there being two locules, one locule of a stamen may fail to develop, or alternatively the two locules may merge late in development to give a single locule. Extreme cases of stamen fusion occur in some species of *Cyclanthera* in the family Cucurbitaceae and in section *Cyclanthera* of genus *Phyllanthus* (family Euphorbiaceae) where the stamens form a ring around the gynoecium, with a single locule.

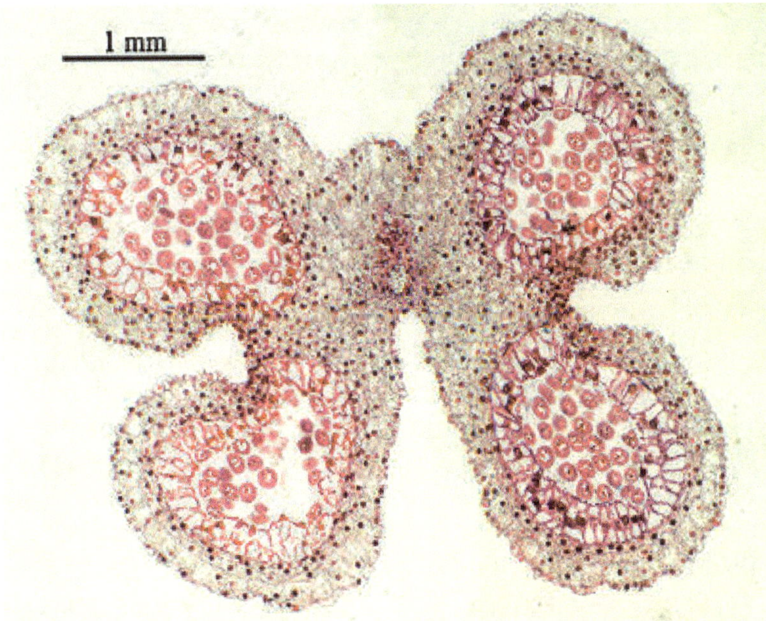

Cross section of a *Lilium* stamen, with four locules surrounded by the tapetum

Pollen Production

A typical anther contains four microsporangia. The *microsporangia* form sacs or pockets (*locules*) in the anther. The two separate locules on each side of an anther may fuse into a single locule. Each microsporangium is lined with a nutritive tissue layer called the *tapetum* and initially contains diploid pollen mother cells. These undergo meiosis to form haploid spores. The spores may remain attached to each other in a tetrad or separate after meiosis. Each microspore then divides mitotically to form an immature microgametophyte called a pollen grain.

The pollen is eventually released when the anther forms openings (dehisces). These may consist of longitudinal slits, pores, as in the heath family (Ericaceae), or by valves, as in the barberry family (Berberidaceae). In some plants, notably members of Orchidaceae and Asclepiadoideae, the pollen remains in masses called pollinia, which are adapted to attach to particular pollinating agents such as birds or insects. More commonly, mature pollen grains separate and are dispensed by wind or water, pollinating insects, birds or other pollination vectors.

Pollen of angiosperms must be transported to the stigma, the receptive surface of the *carpel*, of a compatible flower, for successful pollination to occur. After arriving, the pollen grain (an immature microgametophyte) typically completes its development. It may grow a pollen tube and undergoing mitosis to produce two sperm nuclei.

Sexual Reproduction in Plants

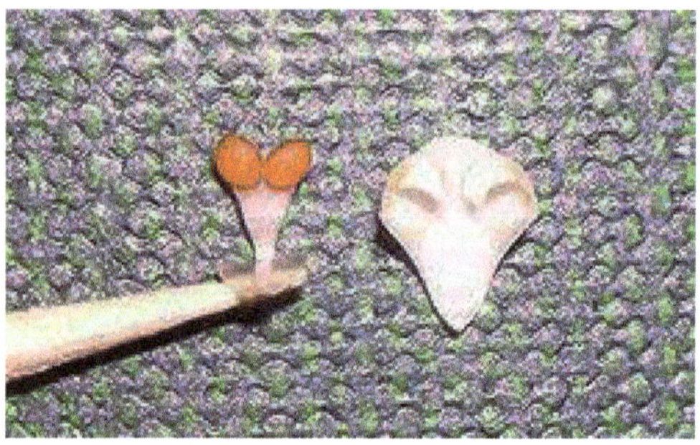

Stamen with pollinia and its anther cap. *Phalaenopsis* orchid.

In the typical flower (that is, in the majority of flowering plant species) each flower has both **carpels** and **stamens**. In some species, however, the flowers are unisexual with only carpels or stamens. (**monoecious** = both types of flowers found on the same plant; **dioecious** = the two types of flower found only on different plants). A flower with only stamens is called **androecious**. A flower with only carpels is called **gynoecious**.

A flower having only functional stamens and lacking functional carpels is called a **staminate flower**, or (inaccurately) male. A plant with only functional carpels is called pistillate, or (inaccurately) female.

An abortive or rudimentary stamen is called a **staminodium** or **staminode**, such as in *Scrophularia nodosa*.

The carpels and stamens of orchids are fused into a column. The top part of the column is formed by the anther, which is covered by an **anther cap**.

Descriptive Terms

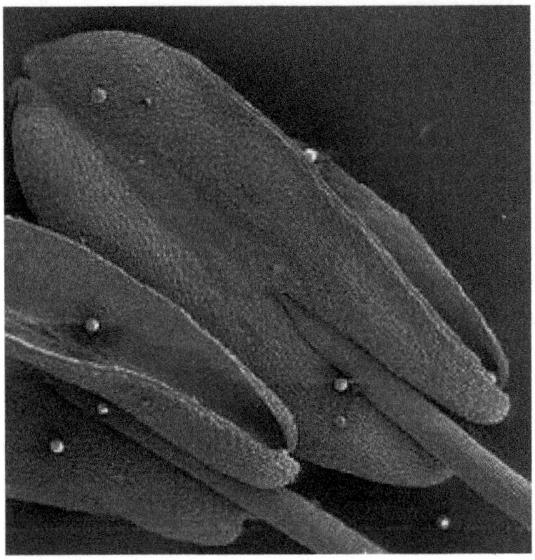

Scanning electron microscope image of *Pentas lanceolata* anthers, with pollen grains on surface

- A column formed from the fusion of multiple filaments is known as an *androphore*.

The anther can be attached to the filament's connective in two ways:

- basifixed: attached at its base to the filament
- pseudobasifixed: a somewhat misnomer configuration where connective tissue extends in a tube around the filament
- dorsifixed: attached at its center to the filament, usually versatile (able to move)

Stamens can be connate (fused or joined in the same whorl):

- extrorse: anther dehiscence directed away from the centre of the flower. Cf. introrse, directed inwards, and latrorse towards the side.
- monadelphous: fused into a single, compound structure
- declinate: curving downwards, then up at the tip (also - declinate-descending)
- diadelphous: joined partially into two androecial structures
- pentadelphous: joined partially into five androecial structures
- synandrous: only the anthers are connate (such as in the Asteraceae). The fused stamens are referred to as a synandrium.

Stamens can also be adnate (fused or joined from more than one whorl):

- epipetalous: adnate to the corolla
- epiphyllous: adnate to undifferentiated tepals (as in many Liliaceae)

They can have different lengths from each other:

- didymous: two equal pairs
- didynamous: occurring in two pairs, a long pair and a shorter pair

- tetradynamous: occurring as a set of six stamens with four long and two shorter ones

or respective to the rest of the flower (perianth):

- exserted: extending beyond the corolla
- included: not extending beyond the corolla

They may be arranged in one of two different patterns:

- spiral; or
- whorled: one or more discrete whorls (series)

They may be arranged, with respect to the petals:

- diplostemonous: in two whorls, the outer alternating with the petals, while the inner is opposite the petals.
- obdiplostemonous: in two whorls, the outer opposite the petals

Column (botany)

The **column**, or technically the **gynostemium**, is a reproductive structure that can be found in several plant families: Aristolochiaceae, Orchidaceae, and Stylidiaceae.

Flower of *Stylidium turbinatum*, showing the column.

It is derived from the fusion of both male and female parts (stamens and pistil) into a single organ. This means that the *style* and *stigma* of the pistil, with the filaments and one or more anthers, are all united.

Orchidaceae

The stigma sits at the apex of the column in the front, but is pointing downwards after **resupination** (the rotation by 180 degrees before unfolding of the flower).

This stigma has the form of a small bowl, the **clinandrium**, a viscous surface embedding the (generally) single anther. On top of it all is the **anther cap**. Sometimes there is a small extension or little beak to the median stigma lobe, called **rostellum**.

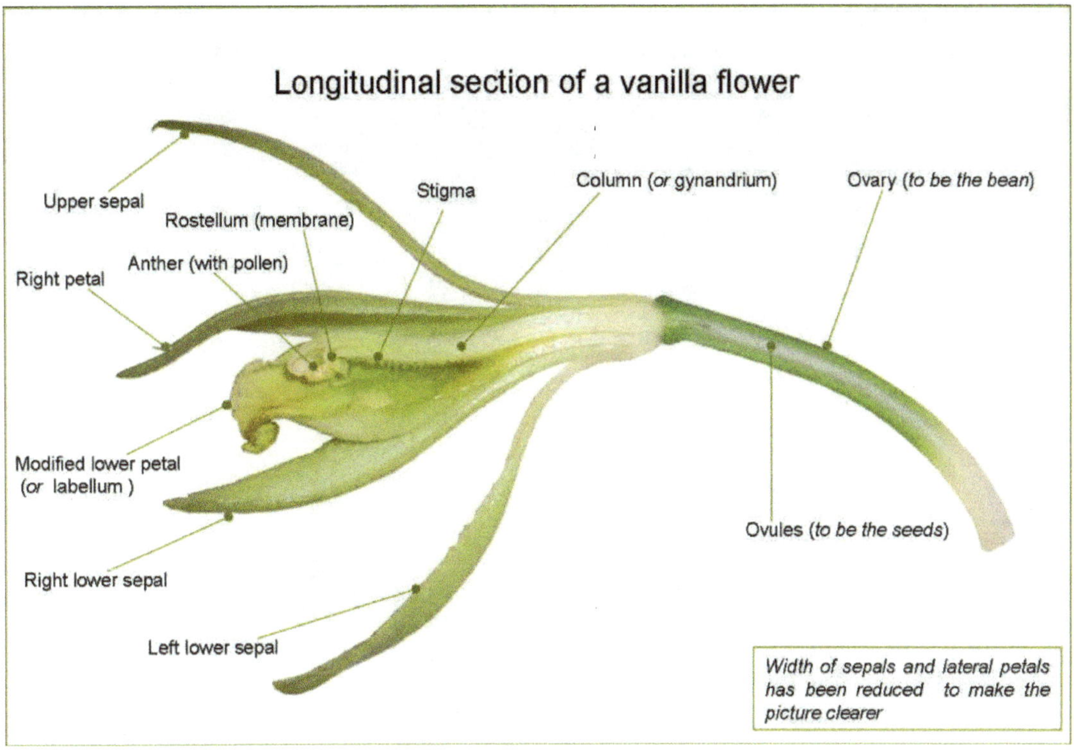

Longitudinal view of a vanilla flower, showing the column

Column wings may project laterally from the stigma. The **column foot** is formed by the attachment of the lip to the basal protruding part of the column. One speaks of a **mentum** (chin) if the lateral sepals are also basally adnate (= attached to the foot of the column).

The column both releases pollen and also receives it (from another individual) for fertilization. In the Orchidaceae family, unlike almost all other flowering plants, the single male anther at the tip of the *column* produces pollen that is not free and powdery but held in waxy masses of two, four or six pellets called ***pollinia***. The transfer of *pollinia* from one flower to another, though highly efficient, is often reliant upon one particular species of arthropods and it can be catastrophic for the population if its pollinator disappears from the community.

SPOROPHYTE

A **sporophyte** is the diploid multicellular stage in the life cycle of a plant or alga. It develops from the zygote produced when a haploid egg cell is fertilized by a haploid sperm and each sporophyte cell therefore has a double set of chromosomes, one set from each parent. All land plants, and most multicellular algae, have life cycles in which a multicellular diploid sporophyte phase alternates with a multicellular haploid gametophyte phase. In the seed plants, the Gymnosperms and flowering plants (*Angiosperms*), the sporophyte phase is more prominent than the gametophyte, and is the familiar green plant with its roots, stem, leaves and cones or flowers. In flowering plants the gametophytes are very reduced in size, and are represented by the pollen and the embryo sac.

Young sporophytes of the common moss *Tortula muralis*. In mosses, the gametophyte is the dominant generation, while the sporophytes consist of sporangium-bearing stalks growing from the tips of the gametophytes

The sporophyte produces spores (hence the name) by meiosis, a process also known as "reduction division" that reduces the number of chromosomes in each spore mother cell by half. The resulting meiospores develop into a gametophyte. Both the spores and the resulting gametophyte are haploid, meaning they only have one set of chromosomes. The mature gametophyte produces male or female gametes (or both) by mitosis. The fusion of male and female gametes produces a diploid zygote which develops into a new sporophyte. This cycle is known as alternation of generations or alternation of phases.

In flowering plants, the sporophyte comprises the whole multicellular body except the

pollen and embryo sac Bryophytes (mosses, liverworts and hornworts) have a dominant gametophyte phase on which the adult sporophyte is dependent for nutrition. The embryo sporophyte develops by cell division of the zygote within the female sex organ or archegonium, and in its early development is therefore nurtured by the gametophyte. Because this embryo-nurturing feature of the life cycle is common to all land plants they are known collectively as the Embryophytes.

Cleistocarpous sporophyte of the moss *Physcomitrella patens*

Most algae have dominant gametophyte generations, but in some species the gametophytes and sporophytes are morphologically similar (isomorphic). An independent sporophyte is the dominant form in all clubmosses, horsetails, ferns, gymnosperms, and angiosperms (flowering plants) that have survived to the present day. Early land plants had sporophytes that produced identical spores (isosporous or homosporous) but the ancestors of the gymnosperms evolved complex heterosporous life cycles in which the spores producing male and female gametophytes were of different sizes, the female megaspores tending to be larger, and fewer in number, than the male microspores.

During the Devonian period several plant groups independently evolved heterospory and subsequently the habit of endospory, in which the gametophytes develop in miniaturized form inside the spore wall. By contrast in exosporous plants, including modern ferns, the gametophytes break the spore wall open on germination and develop outside it. The megagametophytes of endosporic plants such as the seed ferns developed within the sporangia of the parent sporophyte, producing a miniature multicellular female gametophyte complete with female sex organs, or archegonia. The oocytes were fertilized in the archegonia by free-swimming flagellate sperm produced by windborne miniaturized male gametophytes in the form of pre-pollen. The resulting zygote developed into the next sporophyte generation while still retained within the pre-ovule, the single large female meiospore or megaspore contained in the modified sporangium or nucellus of the parent sporophyte. The evolution of heterospory and endospory were among the earliest steps in the evolution of seeds of the kind produced by gymnosperms and angiosperms today.

Plant Embryos

In botany, a seed plant *embryo* is part of a seed, consisting of precursor tissues for the leaves, stem (see **hypocotyl**), and root (see **radicle**), as well as one or more **cotyledons**.

Once the embryo begins to germinate — grow out from the seed — it is called a **seedling** (plantlet).

The inside of a *Ginkgo* seed, showing the embryo

Bryophytes and ferns also produce an embryo, but do not produce seeds. In these plants, the embryo begins its existence attached to the inside of the archegonium on a parental gametophyte from which the egg cell was generated. The inner wall of the archegonium lies in close contact with the "foot" of the developing embryo; this "foot" consists of a bulbous mass of cells at the base of the embryo which may receive nutrition from its parent gametophyte. The structure and development of the rest of the embryo varies by group of plants. Once the embryo has expanded beyond the enclosing archegonium, it is no longer termed an embryo.

FRUIT

In botany, a **fruit** is the seed-bearing structure in flowering plants (also known as angiosperms) formed from the ovary after flowering.

Fruits are the means by which angiosperms disseminate seeds. Edible fruits, in particular, have propagated with the movements of humans and animals in a symbiotic relationship as a means for seed dispersal and nutrition; in fact, humans and many animals have become dependent on fruits as a source of food. Accordingly, fruits account for a substantial fraction of the world's agricultural output, and some (such as the apple and the pomegranate) have acquired extensive cultural and symbolic meanings.

For other uses, see Fruit (disambiguation).

Culinary fruits

Several culinary fruits

Mixed fruit

The Medici citrus collection by Bartolomeo Bimbi, 1715

In common language usage, "fruit" normally means the fleshy seed-associated structures of a plant that are sweet or sour, and edible in the raw state, such as apples, bananas, grapes, lemons, oranges, and strawberries. On the other hand, in botanical usage, "fruit" includes many structures that are not commonly called "fruits", such as bean pods, corn kernels, tomatoes, and wheat grains. The section of a fungus that produces spores is also called a fruiting body.

Botanic fruit and culinary fruit

Venn diagram representing the relationship between (culinary) vegetables and botanical fruits

Many common terms for seeds and fruit do not correspond to the botanical classifications. In culinary terminology, a *fruit* is usually any sweet-tasting plant part, especially a botanical fruit; a *nut* is any hard, oily, and shelled plant product; and a *vegetable* is any savory or less sweet plant product. However, in botany, a *fruit* is the ripened ovary or carpel that contains seeds, a *nut* is a type of fruit and not a seed, and a *seed* is a ripened ovule.

Examples of culinary "vegetables" and nuts that are botanically fruit include corn, cucurbits (e.g., cucumber, pumpkin, and squash), eggplant, legumes (beans, peanuts, and peas), sweet pepper, and tomato. In addition, some spices, such as allspice and chili pepper, are fruits, botanically speaking. In contrast, rhubarb is often referred to as a fruit, because it is used to make sweet desserts such as pies, though only the petiole (leaf stalk) of the rhubarb plant is edible, and edible gymnosperm seeds are often given fruit names, e.g., ginkgo nuts and pine nuts.

Botanically, a cereal grain, such as corn, rice, or wheat, is also a kind of fruit, termed a caryopsis. However, the fruit wall is very thin and is fused to the seed coat, so almost all of the edible grain is actually a seed.

Fruit Structure

The outer, often edible layer, is the *pericarp*, formed from the ovary and surrounding the seeds, although in some species other tissues contribute to or form the edible portion. The pericarp may be described in three layers from outer to inner, the *epicarp*, *mesocarp* and *endocarp*.

Fruit that bears a prominent pointed terminal projection is said to be *beaked*.

Fruit Development

A fruit results from maturation of one or more flowers, and the gynoecium of the flower(s) forms all or part of the fruit.

Inside the ovary/ovaries are one or more ovules where the megagametophyte contains the egg cell. After double fertilization, these ovules will become seeds. The ovules are fertilized in a process that starts with pollination, which involves the movement of pollen from the stamens to the stigma of flowers. After pollination, a tube grows from the pollen through the stigma into the ovary to the ovule and two sperm are transferred from the pollen to the megagametophyte. Within the megagametophyte one of the two sperm unites with the egg, forming a zygote, and the second sperm enters the central cell forming the endosperm mother cell, which completes the double fertilization process. Later the zygote will give rise to the embryo of the seed, and the endosperm mother cell will give rise to endosperm, a nutritive tissue used by the embryo.

The development sequence of a typical drupe, the nectarine (*Prunus persica*) over a 7.5 month period, from bud formation in early winter to fruit ripening in midsummer (see image page for further information)

As the ovules develop into seeds, the ovary begins to ripen and the ovary wall, the *pericarp*, may become fleshy (as in berries or drupes), or form a hard outer covering (as in nuts). In some multiseeded fruits, the extent to which the flesh develops is proportional to the number of fertilized ovules. The pericarp is often differentiated into two or three distinct layers called the *exocarp* (outer layer, also called epicarp), *mesocarp* (middle layer), and *endocarp* (inner layer). In some fruits, especially simple fruits derived from an inferior ovary, other parts of the flower (such as the floral tube, including the petals, sepals, and stamens), fuse with the ovary and ripen with it. In other cases, the sepals, petals and/or stamens and style of the flower fall off. When such other floral parts are a significant part of the fruit, it is called an *accessory fruit*. Since other parts of the flower may contribute to the structure of the fruit, it is important to study flower structure to understand how a particular fruit forms.

There are three general modes of fruit development:

- Apocarpous fruits develop from a single flower having one or more separate carpels, and they are the simplest fruits.
- Syncarpous fruits develop from a single gynoecium having two or more carpels fused together.
- Multiple fruits form from many different flowers.

Plant scientists have grouped fruits into three main groups, simple fruits, aggregate fruits, and composite or multiple fruits.The groupings are not evolutionarily relevant, since many diverse plant taxa may be in the same group, but reflect how the flower organs are arranged and how the fruits develop.

Simple fruit

Simple fruits can be either dry or fleshy, and result from the ripening of a simple or compound ovary in a flower with only one pistil. Dry fruits may be either dehiscent (they open to discharge seeds), or indehiscent (they do not open to discharge seeds). Types of dry, simple fruits, and examples of each, include:

Epigynous berries are simple fleshy fruit. Clockwise from top right: cranberries, lingonberries, blueberries, red huckleberries

- achene – Most commonly seen in aggregate fruits (e.g., strawberry)
- capsule – (e.g., Brazil nut)
- caryopsis – (e.g., wheat)
- cypsela – an achene-like fruit derived from the individual florets in a capitulum (e.g., dandelion).
- fibrous drupe – (e.g., coconut, walnut)
- follicle – is formed from a single carpel, opens by one suture (e.g., milkweed), commonly seen in aggregate fruits (e.g., magnolia)
- legume – (e.g., bean, pea, peanut)
- loment – a type of indehiscent legume
- nut – (e.g., beech, hazelnut, oak acorn)
- samara – (e.g., ash, elm, maple key)
- schizocarp – (e.g., carrot seed)
- silique – (e.g., radish seed)
- silicle – (e.g., shepherd's purse)

- utricle – (e.g., beet)

Lilium unripe capsule fruit

Fruits in which part or all of the *pericarp* (fruit wall) is fleshy at maturity are *simple fleshy fruits*. Types of simple, fleshy, fruits (with examples) include:

- berry – (e.g., cranberry, gooseberry, redcurrant, tomato)
- stone fruit or drupe (e.g., apricot, cherry, olive, peach, plum)

An aggregate fruit, or *etaerio*, develops from a single flower with numerous simple pistils.

- Magnolia and peony, collection of follicles developing from one flower.
- Sweet gum, collection of capsules.
- Sycamore, collection of achenes.
- Teasel, collection of cypsellas
- Tuliptree, collection of samaras.

The pome fruits of the family Rosaceae, (including apples, pears, rosehips, and saskatoon berry) are a syncarpous fleshy fruit, a simple fruit, developing from a half-inferior ovary.

Dewberry flowers. Note the multiple pistils, each of which will produce a drupelet. Each flower will become a blackberry-like aggregate fruit.

Schizocarp fruits form from a syncarpous ovary and do not really dehisce, but rather split into segments with one or more seeds; they include a number of different forms from a wide range of families. Carrot seed is an example.

Aggregate fruit

Aggregate fruits form from single flowers that have multiple carpels which are not joined together, i.e. each pistil contains one carpel. Each pistil forms a fruitlet, and collectively the fruitlets are called an etaerio. Four types of aggregate fruits include etaerios of achenes, follicles, drupelets, and berries. Ranunculaceae species, including *Clematis* and *Ranunculus* have an etaerio of achenes, *Calotropis* has an etaerio of follicles, and *Rubus* species like raspberry, have an etaerio of drupelets. *Annona* have an etaerio of berries.

Detail of raspberry flower

The raspberry, whose pistils are termed *drupelets* because each is like a small drupe attached to the receptacle. In some bramble fruits (such as blackberry) the receptacle is elongated and part of the ripe fruit, making the blackberry an *aggregate-accessory* fruit. The strawberry is also an aggregate-accessory fruit, only one in which the seeds are contained in achenes. In all these examples, the fruit develops from a single flower with numerous pistils.

Multiple Fruits

A multiple fruit is one formed from a cluster of flowers (called an *inflorescence*). Each flower produces a fruit, but these mature into a single mass. Examples are the pineapple, fig, mulberry, osage-orange, and breadfruit.

In some plants, such as this noni, flowers are produced regularly along the stem and it is possible to see together examples of flowering, fruit development, and fruit ripening.

In the photograph on the right, stages of flowering and fruit development in the noni or Indian mulberry (*Morinda citrifolia*) can be observed on a single branch. First an inflorescence of white flowers called a head is produced. After fertilization, each flower develops into a drupe, and as the drupes expand, they become *connate* (merge) into a *multiple fleshy fruit* called a *syncarp*.

Berries

Berries are another type of fleshy fruit; they are simple fruit created from a single ovary. The ovary may be compound, with several carpels. Types include (examples follow in the table below):

- Pepo – berries whose skin is hardened, cucurbits
- Hesperidium – berries with a rind and a juicy interior, like most citrus fruit

Accessory fruit

Some or all of the edible part of accessory fruit is not generated by the ovary. Accessory fruit can be simple, aggregate, or multiple, i.e., they can include one or more pistils and other parts from the same flower, or the pistils and other parts of many flowers.

The fruit of a pineapple includes tissue from the sepals as well as the pistils of many flowers. It is an accessory fruit and a multiple fruit.

Table of fruit examples

Types of fleshy fruits					
True berry	**Pepo**	**Hesperidium**	**Aggregate fruit**	**Multiple fruit**	**Accessory fruit**
Blackcurrant, Blueberry, Chili pepper, Cranberry, Eggplant, Gooseberry, Grape, Guava, Kiwifruit, Lucuma, Pomegranate, Redcurrant, Tomato	Cucumber, Gourd, Melon, Pumpkin	Grapefruit, Lemon, Lime, Orange	Blackberry, Boysenberry, Raspberry	Fig, Hedge apple, Mulberry, Pineapple	Apple, Pineapple, Rose hip, Stone fruit, Strawberry

Seedless Fruits

Seedlessness is an important feature of some fruits of commerce. Commercial cultivars of bananas and pineapples are examples of seedless fruits. Some cultivars of citrus fruits (especially grapefruit, mandarin oranges, navel oranges), satsumas, table grapes, and watermelons are valued for their seedlessness. In some species, seedlessness is the result of *parthenocarpy*, where fruits set without fertilization. Parthenocarpic fruit set may or may not require pollination, but most seedless citrus fruits require a stimulus from pollination to produce fruit.

Seedless bananas and grapes are triploids, and seedlessness results from the abortion of the embryonic plant that is produced by fertilization, a phenomenon known as *stenospermocarpy*, which requires normal pollination and fertilization.

An arrangement of fruits commonly thought of as vegetables, including tomatoes and various squash

Seed Dissemination

Variations in fruit structures largely depend on their seeds' mode of dispersal. This dispersal can be achieved by animals, explosive dehiscence, water, or wind.

Grapes and Mangoes

Some fruits have coats covered with spikes or hooked burrs, either to prevent themselves from being eaten by animals, or to stick to the feathers, hairs, or legs of animals, using them as dispersal agents. Examples include cocklebur and unicorn plant.

The sweet flesh of many fruits is "deliberately" appealing to animals, so that the seeds held within are eaten and "unwittingly" carried away and deposited (i.e., defecated) at a distance from the parent. Likewise, the nutritious, oily kernels of nuts are appealing to rodents (such as squirrels), which hoard them in the soil to avoid starving during the winter, thus giving those seeds that remain uneaten the chance to germinate and grow into a new plant away from their parent.

Other fruits are elongated and flattened out naturally, and so become thin, like wings or helicopter blades, e.g., elm, maple, and tuliptree. This is an evolutionary mechanism to increase dispersal distance away from the parent, via wind. Other wind-dispersed fruit have tiny "parachutes", e.g., dandelion, milkweed, salsify.

Coconut fruits can float thousands of miles in the ocean to spread seeds. Some other fruits that can disperse via water are nipa palm and screw pine.

Some fruits fling seeds substantial distances (up to 100 m in sandbox tree) via explosive dehiscence or other mechanisms, e.g., impatiens and squirting cucumber.

Uses

Nectarines are one of many fruits that can be easily stewed.

Oranges, bananas, pears, apples, and a watermelon

Many hundreds of fruits, including fleshy fruits (like apple, kiwifruit, mango, peach, pear, and watermelon) are commercially valuable as human food, eaten both fresh and as jams, marmalade and other preserves. Fruits are also used in manufactured foods (e.g., cakes, cookies, ice cream, muffins, or yogurt) or beverages, such as fruit juices (e.g., apple juice, grape juice, or orange juice) or alcoholic beverages (e.g., brandy, fruit beer, or wine), Fruits are also used for gift giving, e.g., in the form of Fruit Baskets and Fruit Bouquets.

Many "vegetables" in culinary *parlance* are botanical fruits, including bell pepper, cucumber, eggplant, green bean, okra, pumpkin, squash, tomato, and zucchini.Olive fruit is pressed for olive oil. Spices like allspice, black pepper, paprika, and vanilla are derived from berries.

Nutritional Value

Each point refers to a 100 g serving of the fresh fruit, the daily recommended allowance of vitamin C is on the X axis and mg of Potassium (K) on the Y (offset by 100 mg which every fruit has) and the size of the disk represents amount of fiber (key in upper right). Watermelon, which has almost no fiber, and low levels of vitamin C and potassium, comes in last place.

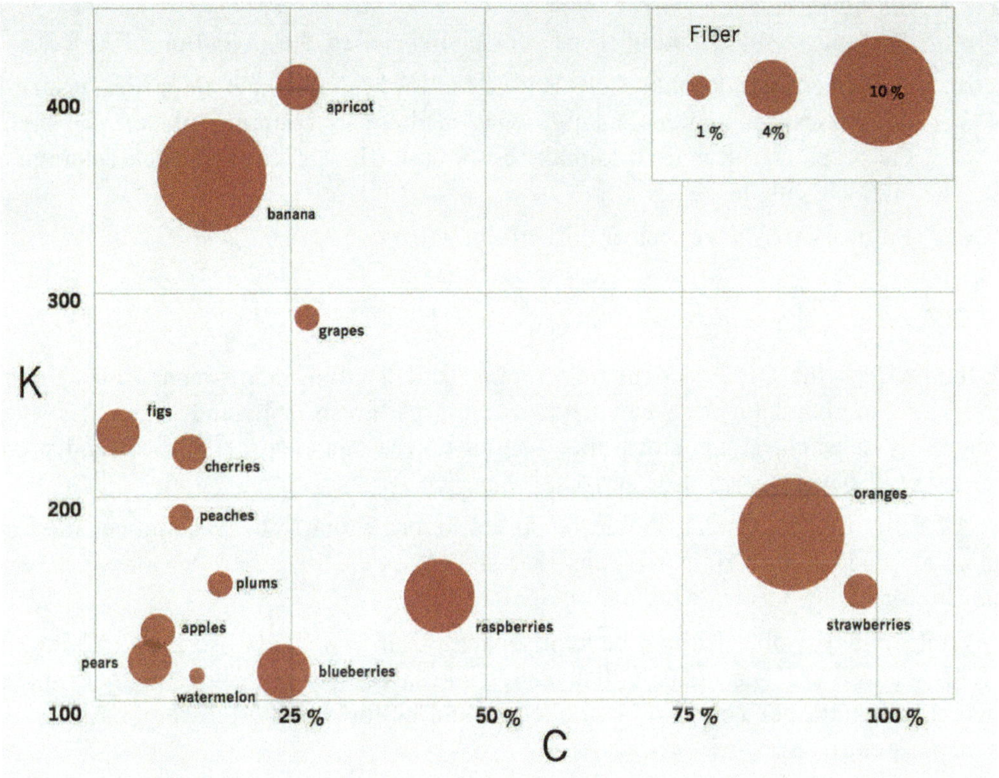

Fresh fruits are generally high in fiber, vitamin C, and water.

Regular consumption of fruit is generally associated with reduced risks of several diseases and functional declines associated with aging.

Nonfood uses

Because fruits have been such a major part of the human diet, various cultures have developed many different uses for fruits they do not depend on for food. For example:

- Bayberry fruits provide a wax often used to make candles;
- Many dry fruits are used as decorations or in dried flower arrangements (e.g., annual honesty, cotoneaster, lotus, milkweed, unicorn plant, and wheat). Ornamental trees and shrubs are often cultivated for their colorful fruits, including beautyberry, cotoneaster, holly, pyracantha, skimmia, and viburnum.
- Fruits of opium poppy are the source of opium, which contains the drugs codeine and morphine, as well as the biologically inactive chemical theabaine from which the drug oxycodone is synthesized.
- Osage orange fruits are used to repel cockroaches.
- Many fruits provide natural dyes (e.g., cherry, mulberry, sumac, and walnut).

- Dried gourds are used as bird houses, cups, decorations, dishes, musical instruments, and water jugs.
- Pumpkins are carved into Jack-o'-lanterns for Halloween.
- The spiny fruit of burdock or cocklebur inspired the invention of Velcro.
- Coir fiber from coconut shells is used for brushes, doormats, floor tiles, insulation, mattresses, sacking, and as a growing medium for container plants. The shell of the coconut fruit is used to make bird houses, bowls, cups, musical instruments, and souvenir heads.
- Fruit is often a subject of still life paintings.

Safety

For food safety, the CDC recommends proper fruit handling and preparation to reduce the risk of food contamination and foodborne illness. Fresh fruits and vegetables should be carefully selected; at the store, they should not be damaged or bruised; and pre-cut pieces should be refrigerated or surrounded by ice.

All fruits and vegetables should be rinsed before eating. This recommendation also applies to produce with rinds or skins that are not eaten. It should be done just before preparing or eating to avoid premature spoilage.

Fruits and vegetables should be kept separate from raw foods like meat, poultry, and seafood, as well as from utensils that have come in contact with raw foods. Fruits and vegetables that are not going to be cooked should be thrown away if they have touched raw meat, poultry, seafood, or eggs.

All cut, peeled, or cooked fruits and vegetables should be refrigerated within two hours. After a certain time, harmful bacteria may grow on them and increase the risk of foodborne illness.

Allergies

Fruit allergies make up about 10 percent of all food related allergies

Storage

All fruits benefit from proper post harvest care, and in many fruits, the plant hormone ethylene causes ripening. Therefore, maintaining most fruits in an efficient cold chain is optimal for post harvest storage, with the aim of extending and ensuring shelf life.

REFERENCES

1. Beentje, Henk (2010). The Kew Plant Glossary. Richmond, Surrey: Royal Botanic Gardens, Kew. ISBN 978-1-84246-422-9., p. 10
2. Sattler, R. 1973. Organogenesis of Flowers. A Photographic Text-Atlas. University of Toronto Press. ISBN 0-8020-1864-5.

3. Sattler, R. 1988. A dynamic multidimensional approach to floral morphology. In: Leins, P., Tucker, S. C. and Endress, P. (eds) Aspects of Floral Development. J. Cramer, Berlin, pp. 1-6. ISBN 3-443-50011-0
4. Greyson, R. I. 1994. The Development of Flowers. Oxford University Press. ISBN 0-19-506688-X.
5. Leins, P. and Erbar, C. 2010. Flower and Fruit. Schweizerbart Science Publishers, Stuttgart. ISBN 978-3-510-65261-7.
6. Lewis, C.T. & Short, C. (1879). A Latin dictionary founded on Andrews' edition of Freund's Latin dictionary. Oxford: Clarendon Press.
7. Klein, E. (1971). A comprehensive etymological dictionary of the English language. Dealing with the origin of words and their sense development thus illustration the history of civilization and culture. Amsterdam: Elsevier Science B.V.
8. Siebenhaar, F.J. (1850). Terminologisches Wörterbuch der medicinischen Wissenschaften. (Zweite Auflage). Leipzig: Arnoldische Buchhandlung.
9. Saalfeld, G.A.E.A. (1884). Tensaurus Italograecus. Ausführliches historisch-kritisches Wörterbuch der Griechischen Lehn- und Fremdwörter im Lateinischen. Wien: Druck und Verlag von Carl Gerold's Sohn, Buchhändler der Kaiserl. Akademie der Wissenschaften.
10. Liddell, H.G. & Scott, R. (1940). A Greek-English Lexicon. revised and augmented throughout by Sir Henry Stuart Jones. with the assistance of. Roderick McKenzie. Oxford: Clarendon Press.
11. Goebel, K.E.v. (1969) [1905]. Organography of plants, especially of the Archegoniatae and Spermaphyta. Part 2 Special organography. New York: Hofner publishing company. pages 553 555
12. Rendle, A.B. (1925). The Classification of Flowering Plants. Cambridge University Press. ISBN 9780521060578.
13. Encyclopædia Britannica.com
14. Hickey, M.; King, C. (1997). Common Families of Flowering Plants. Cambridge University Press. ISBN 9780521576093.
15. William G. D'Arcy, Richard C. Keating (eds.) The Anther: Form, Function, and Phylogeny. Cambridge University Press, 1996 ISBN 0521480639, 9780521480635

CHAPTER 5

Plant Physiology

Chapter Contents

Plant Physiology and Biochemistry publishes original theoretical, experimental and technical contributions in the various fields of plant physiology (biochemistry, physiology, structure, genetics, plant-microbe interactions, etc.) at diverse levels of integration (molecular, subcellular, cellular, organ, whole plant, environmental).

INTRODUCTION

Plant physiology is a subdiscipline of botany concerned with the functioning, or physiology, of plants. Closely related fields include plant morphology (structure of plants), plant ecology (interactions with the environment), phytochemistry (biochemistry of plants), cell biology, genetics, biophysics and molecular biology.

Fundamental processes such as photosynthesis, respiration, plant nutrition, plant hormone functions, tropisms, nastic movements, photoperiodism, photomorphogenesis, circadian rhythms, environmental stress physiology, seed germination, dormancy and stomata function and transpiration, both parts of plant water relations, are studied by plant physiologists.

A germination rate experiment

AIMS

The field of plant physiology includes the study of all the internal activities of plants—those chemical and physical processes associated with life as they occur in plants. This includes study at many levels of scale of size and time. At the smallest scale are molecular interactions of photosynthesis and internal diffusion of water, minerals, and nutrients. At the largest scale are the processes of plant development, seasonality, dormancy, and reproductive control. Major subdisciplines of plant physiology include phytochemistry (the study of the biochemistry of plants) and phytopathology (the study of disease in plants). The scope of plant physiology as a discipline may be divided into several major areas of research.

First, the study of phytochemistry (plant chemistry) is included within the domain of plant physiology. To function and survive, plants produce a wide array of chemical compounds not found in other organisms. Photosynthesis requires a large array of pigments, enzymes, and other compounds to function. Because they cannot move, plants must also defend themselves chemically from herbivores, pathogens and competition from other plants. They do this by producing toxins and foul-tasting or smelling chemicals. Other compounds defend plants against disease, permit survival during drought, and prepare plants for dormancy, while other compounds are used to attract pollinators or herbivores to spread ripe seeds.

Secondly, plant physiology includes the study of biological and chemical processes of individual plant cells. Plant cells have a number of features that distinguish them from cells of animals, and which lead to major differences in the way that plant life behaves and responds differently from animal life. For example, plant cells have a cell wall which restricts the shape of plant cells and thereby limits the flexibility and mobility of plants. Plant cells also contain chlorophyll, a chemical compound that interacts with light in a way that enables plants to manufacture their own nutrients rather than consuming other living things as animals do.

Thirdly, plant physiology deals with interactions between cells, tissues, and organs within a plant. Different cells and tissues are physically and chemically specialized to perform different functions. Roots and rhizoids function to anchor the plant and acquire minerals in the soil. Leaves catch light in order to manufacture nutrients. For both of these organs to remain living, minerals that the roots acquire must be transported to the leaves, and the nutrients manufactured in the leaves must be transported to the roots. Plants have developed a number of ways to achieve this transport, such as vascular tissue, and the functioning of the various modes of transport is studied by plant physiologists.

Fourthly, plant physiologists study the ways that plants control or regulate internal functions. Like animals, plants produce chemicals called hormones which are produced in one part of the plant to signal cells in another part of the plant to respond. Many flowering plants bloom at the appropriate time because of light-sensitive compounds that respond to the length of the night, a phenomenon known as photoperiodism. The ripening of fruit and loss of leaves in the winter are controlled in part by the production of the gas ethylene by the plant.

Finally, plant physiology includes the study of plant response to environmental conditions and their variation, a field known as environmental physiology. Stress from water loss, changes in air chemistry, or crowding by other plants can lead to changes in the way a plant functions. These changes may be affected by genetic, chemical, and physical factors.

BIOCHEMISTRY OF PLANTS

The chemical elements of which plants are constructed—principally carbon, oxygen, hydrogen, nitrogen, phosphorus, sulfur, etc.—are the same as for all other life forms animals, fungi, bacteria and even viruses. Only the details of the molecules into which they are assembled differs.

Despite this underlying similarity, plants produce a vast array of chemical compounds with unique properties which they use to cope with their environment. Pigments are used by plants to absorb or detect light, and are extracted by humans for use in dyes. Other plant products may be used for the manufacture of commercially important rubber or biofuel. Perhaps the most celebrated compounds from plants are those with pharmacological activity, such as salicylic acid from which aspirin is made, morphine, and digoxin. Drug companies spend billions of dollars each year researching plant compounds for potential medicinal benefits.

Latex being collected from a tapped rubber tree.

Constituent Elements

Plants require some nutrients, such as carbon and nitrogen, in large quantities to survive. Such nutrients are termed macronutrients, where the prefix *macro-* (large) refers to the quantity needed, not the size of the nutrient particles themselves. Other nutrients, called micronutrients, are required only in trace amounts for plants to remain healthy. Such micronutrients are usually absorbed as ions dissolved in water taken from the soil, though carnivorous plants acquire some of their micronutrients from captured prey.

The following tables list element nutrients essential to plants. Uses within plants are generalized.

Macronutrients – necessary in large quantities

Element	Form of uptake	Notes
Nitrogen	NO_3^-, NH_4^+	Nucleic acids, proteins, hormones, etc.
Oxygen	O_2 H_2O	Cellulose, starch, other organic compounds
Carbon	CO_2	Cellulose, starch, other organic compounds
Hydrogen	H_2O	Cellulose, starch, other organic compounds
Potassium	K^+	Cofactor in protein synthesis, water balance, etc.
Calcium	Ca^{2+}	Membrane synthesis and stabilization
Magnesium	Mg^{2+}	Element essential for chlorophyll
Phosphorus	$H_2PO_4^-$	Nucleic acids, phospholipids, ATP
Sulfur	SO_4^{2-}	Constituent of proteins

Micronutrients – necessary in small quantities

Element	Form of uptake	Notes
Chlorine	Cl^-	Photosystem II and stomata function
Iron	Fe^{2+}, Fe^{3+}	Chlorophyll formation
Boron	HBO_3	Crosslinking pectin
Manganese	Mn^{2+}	Activity of some enzymes
Zinc	Zn^{2+}	Involved in the synthesis of enzymes and chlorophyll
Copper	Cu^+	Enzymes for lignin synthesis

Molybdenum	MoO_4^{2-}	Nitrogen fixation, reduction of nitrates
Nickel	Ni^{2+}	Enzymatic cofactor in the metabolism of nitrogen compounds

Pigments

Among the most important molecules for plant function are the pigments. Plant pigments include a variety of different kinds of molecules, including porphyrins, carotenoids, and anthocyanins. All biological pigments selectively absorb certain wavelengths of light while reflecting others. The light that is absorbed may be used by the plant to power chemical reactions, while the reflected wavelengths of light determine the color the pigment appears to the eye. Chlorophyll is the primary pigment in plants; it is a porphyrin that absorbs red and blue wavelengths of light while reflecting green. It is the presence and relative abundance of chlorophyll that gives plants their green color. All land plants and green algae possess two forms of this pigment: chlorophyll *a* and chlorophyll *b*. Kelps, diatoms, and other photosynthetic heterokonts contain chlorophyll *c* instead of *b*, red algae possess chlorophyll *a* and « d". All chlorophylls serve as the primary means plants use to intercept light to fuel photosynthesis.

Space-filling model of the chlorophyll molecule.

Anthocyanin gives these pansies their dark purple pigmentation.

Carotenoids are red, orange, or yellow tetraterpenoids. They function as accessory pigments in plants, helping to fuel photosynthesis by gathering wavelengths of light not readily absorbed by chlorophyll. The most familiar carotenoids are carotene (an orange pigment found in carrots), lutein (a yellow pigment found in fruits and vegetables), and lycopene (the red pigment responsible for the color of tomatoes). Carotenoids have been shown to act as antioxidants and to promote healthy eyesight in humans.

Anthocyanins (literally "flower blue") are water-soluble flavonoid pigments that appear red to blue, according to pH. They occur in all tissues of higher plants, providing color in leaves, stems, roots, flowers, and fruits, though not always in sufficient quantities to be noticeable. Anthocyanins are most visible in the petals of flowers, where they may make up as much as 30% of the dry weight of the tissue. They are also responsible for the purple color seen on the underside of tropical shade plants such as *Tradescantia zebrina*. In these plants, the anthocyanin catches light that has passed through the leaf and reflects it back towards regions bearing chlorophyll, in order to maximize the use of available light

Betalains are red or yellow pigments. Like anthocyanins they are water-soluble, but unlike anthocyanins they are indole-derived compounds synthesized from tyrosine. This class of pigments is found only in the Caryophyllales (including cactus and amaranth), and never co-occur in plants with anthocyanins. Betalains are responsible for the deep red color of beets, and are used commercially as food-coloring agents. Plant physiologists are uncertain of the function that betalains have in plants which possess them, but there is some preliminary evidence that they may have fungicidal properties.

Signals and Regulators

A mutation that stops *Arabidopsis thaliana* responding to auxin causes abnormal growth (right)

Plants produce hormones and other growth regulators which act to signal a physiological response in their tissues. They also produce compounds such as phytochrome that are sensitive to light and which serve to trigger growth or development in response to environmental signals.

Plant Hormones

Plant hormones, known as plant growth regulators (PGRs) or phytohormones, are chemicals that regulate a plant's growth. According to a standard animal definition, hormones are signal molecules produced at specific locations, that occur in very low concentrations, and cause altered processes in target cells at other locations. Unlike animals, plants lack specific hormone-producing tissues or organs. Plant hormones are often not transported to other parts of the plant and production is not limited to specific locations.

Plant hormones are chemicals that in small amounts promote and influence the growth, development and differentiation of cells and tissues. Hormones are vital to plant growth; affecting processes in plants from flowering to seed development, dormancy, and germination. They regulate which tissues grow upwards and which grow downwards, leaf formation and stem growth, fruit development and ripening, as well as leaf abscission and even plant death.

The most important plant hormones are abscissic acid (ABA), auxins, ethylene, gibberellins, and cytokinins, though there are many other substances that serve to regulate plant physiology.

Photomorphogenesis

While most people know that light is important for photosynthesis in plants, few realize that plant sensitivity to light plays a role in the control of plant structural development (morphogenesis). The use of light to control structural development is called photomorphogenesis, and is dependent upon the presence of specialized photoreceptors, which are chemical pigments capable of absorbing specific wavelengths of light.

Plants use four kinds of photoreceptors: phytochrome, cryptochrome, a UV-B photoreceptor, and protochlorophyllide *a*. The first two of these, phytochrome and cryptochrome, are photoreceptor proteins, complex molecular structures formed by joining a protein with a light-sensitive pigment. Cryptochrome is also known as the UV-A photoreceptor, because it absorbs ultraviolet light in the long wave "A" region. The UV-B receptor is one or more compounds not yet identified with certainty, though some evidence suggests carotene or riboflavin as candidates. Protochlorophyllide *a*, as its name suggests, is a chemical precursor of chlorophyll.

The most studied of the photoreceptors in plants is phytochrome. It is sensitive to light in the red and far-red region of the visible spectrum. Many flowering plants use it to regulate the time of flowering based on the length of day and night (photoperiodism) and to set circadian rhythms. It also regulates other responses including the germination of seeds, elongation of seedlings, the size, shape and number of leaves, the synthesis of chlorophyll, and the straightening of the epicotyl or hypocotyl hook of dicot seedlings.

Photoperiodism

The poinsettia is a short-day plant, requiring two months of long nights prior to blooming.

Many flowering plants use the pigment phytochrome to sense seasonal changes in day length, which they take as signals to flower. This sensitivity to day length is termed photoperiodism. Broadly speaking, flowering plants can be classified as long day plants, short day plants, or day neutral plants, depending on their particular response to changes in day length. Long day plants require a certain minimum length of daylight to starts flowering, so these plants flower in the spring or summer. Conversely, short day plants flower when the length of daylight falls below a certain critical level. Day neutral plants do not initiate flowering based on photoperiodism, though some may use temperature sensitivity (vernalization) instead. Although a short day plant cannot flower during the long days of summer, it is not actually the period of light exposure that limits flowering. Rather, a short day plant requires a minimal length of uninterrupted darkness in each 24-hour period (a short daylength) before floral development can begin. It has been determined experimentally that a short day plant (long night) does not flower if a flash of phytochrome activating light is used on the plant during the night. Plants make use of the phytochrome system to sense day length or photoperiod. This fact is utilized by florists and greenhouse gardeners to control and even induce flowering out of season, such as the *Poinsettia*.

Environmental Physiology

Paradoxically, the subdiscipline of environmental physiology is on the one hand a recent field of study in plant ecology and on the other hand one of the oldest.Environmental physiology is the preferred name of the subdiscipline among plant physiologists, but it goes by a number of other names in the applied sciences. It is roughly synonymous with ecophysiology, crop ecology, horticulture and agronomy. The particular name applied to the subdiscipline is specific to the viewpoint and goals of research. Whatever name is applied, it deals with the ways in which plants respond to their environment and so overlaps with the field of ecology. Environmental physiologists examine plant response to physical factors such as radiation (including light and ultraviolet radiation), temperature, fire, and wind. Of particular importance are water relations (which can be measured with the Pressure bomb) and the stress of drought or inundation, exchange of gases with the atmosphere, as well as the cycling of nutrients such as nitrogen and carbon.

Phototropism in *Arabidopsis thaliana* is regulated by blue to UV light.

Environmental physiologists also examine plant response to biological factors. This includes not only negative interactions, such as competition, herbivory, disease and parasitism, but also positive interactions, such as mutualism and pollination.

Tropisms and Nastic Movements

Plants may respond both to directional and non-directional stimuli. A response to a directional stimulus, such as gravity or sunlight, is called a tropism. A response to a nondirectional stimulus, such as temperature or humidity, is a nastic movement.

Tropisms in plants are the result of differential cell growth, in which the cells on one side of the plant elongates more than those on the other side, causing the part to bend toward the side with less growth. Among the common tropisms seen in plants is phototropism, the bending of the plant toward a source of light. Phototropism allows the plant to maximize light exposure in plants which require additional light for photosynthesis, or to minimize it in plants subjected to intense light and heat. Geotropism allows the roots of a plant to determine the direction of gravity and grow downwards. Tropisms generally result from an interaction between the environment and production of one or more plant hormones.

Nastic movements results from differential cell growth (e.g. epinasty and hiponasty), or from changes in turgor pressure within plant tissues (e.g., nyctinasty), which may occur rapidly. A familiar example is thigmonasty (response to touch) in the Venus fly trap, a carnivorous plant. The traps consist of modified leaf blades which bear sensitive trigger hairs. When the hairs are touched by an insect or other animal, the leaf folds shut. This mechanism allows the plant to trap and digest small insects for additional nutrients.

Although the trap is rapidly shut by changes in internal cell pressures, the leaf must grow slowly to reset for a second opportunity to trap insects.

Plant Disease

Powdery mildew on crop leaves

Economically, one of the most important areas of research in environmental physiology is that of phytopathology, the study of diseases in plants and the manner in which plants resist or cope with infection. Plant are susceptible to the same kinds of disease organisms as animals, including viruses, bacteria, and fungi, as well as physical invasion by insects and roundworms. Because the biology of plants differs with animals, their symptoms and responses are quite different. In some cases, a plant can simply shed infected leaves or flowers to prevent the spread of disease, in a process called abscission. Most animals do not have this option as a means of controlling disease. Plant diseases organisms themselves also differ from those causing disease in animals because plants cannot usually spread infection through casual physical contact. Plant pathogens tend to spread via spores or are carried by animal vectors. One of the most important advances in the control of plant disease was the discovery of Bordeaux mixture in the nineteenth century. The mixture is the first known fungicide and is a combination of copper sulfate and lime. Application of the mixture served to inhibit the growth of downy mildew that threatened to seriously damage the French wine industry.

HISTORY

Early History

Sir Francis Bacon published one of the first plant physiology experiments in 1627 in the book, *Sylva Sylvarum.* Bacon grew several terrestrial plants, including a rose, in water and concluded that soil was only needed to keep the plant upright. Jan Baptist van Helmont published what is considered the first quantitative experiment in plant physiology in 1648. He grew a willow tree for five years in a pot containing 200 pounds of oven-dry soil. The soil lost just two ounces of dry weight and van Helmont concluded that plants get all their weight from water, not soil. In 1699, John Woodward published experiments on growth of spearmint in different sources of water. He found that plants grew much better in water with soil added than in distilled water.

Jan Baptist van Helmont.

Stephen Hales is considered the Father of Plant Physiology for the many experiments in the 1727 book; though Julius von Sachs unified the pieces of plant physiology and put them together as a discipline. His *Lehrbuch der Botanik* was the plant physiology bible of its time.

Researchers discovered in the 1800s that plants absorb essential mineral nutrients as inorganic ions in water. In natural conditions, soil acts as a mineral nutrient reservoir but the soil itself is not essential to plant growth. When the mineral nutrients in the soil are dissolved in water, plant roots absorb nutrients readily, soil is no longer required for the plant to thrive. This observation is the basis for hydroponics, the growing of plants in a water solution rather than soil, which has become a standard technique in biological research, teaching lab exercises, crop production and as a hobby.

Current Research

One of the leading journals in the field is *Plant Physiology*, started in 1926. All its back issues are available online for free.Many other journals often carry plant physiology articles, including *Physiologia Plantarum*, *Journal of Experimental Botany*, *American Journal of Botany*, *Annals of Botany*, *Journal of Plant Nutrition* and *Proceedings of the National Academy of Sciences.*

Economic Applications

Food Production

In horticulture and agriculture along with food science, plant physiology is an important topic relating to fruits, vegetables, and other consumable parts of plants. Topics studied include: *climatic* requirements, fruit drop, nutrition, ripening, fruit set. The production of food crops also hinges on the study of plant physiology covering such topics as optimal planting and harvesting times and post harvest storage of plant products for human consumption and the production of secondary products like drugs and cosmetics.

PLANT NUTRITION

Plant nutrition is the study of the chemical elements and compounds necessary for plant growth, plant metabolism and their external supply. In 1972, E. Epstein defined two criteria for an element to be essential for plant growth:

- in its absence the plant is unable to complete a normal life cycle.
- or that the element is part of some essential plant constituent or metabolite.

This is in accordance with Justus von Liebig's law of the minimum. The essential plant nutrients include carbon and oxygen which are absorbed from the air, whereas other nutrients including hydrogen are typically obtained from the soil (exceptions include some parasitic or carnivorous plants).

Plants must obtain the following mineral nutrients from their growing medium:

- the macronutrients: nitrogen (N), phosphorus (P), potassium (K), calcium (Ca), sulfur (S), magnesium (Mg); and
- the micronutrients (or trace minerals): boron (B), chlorine (Cl), manganese (Mn), iron (Fe), zinc (Zn), copper (Cu), molybdenum (Mo), nickel (Ni).

The macronutrients are consumed in larger quantities and are usually present in plant tissue in concentrations of between 0.2% and 4.0% (on a dry matter weight basis). Micronutrients are present in plant tissue in quantities measured in parts per million, ranging from 0.1 to 200 ppm, or less than 0.02% dry weight.

Most soil conditions across the world can provide plants adapted to that climate and soil with sufficient nutrition for a complete life cycle, without the addition of nutrients as fertilizer. However, if the soil is cropped it is necessary to artificially modify soil fertility through the addition of fertilizer to promote vigorous growth and increase or sustain yield. This is done because, even with adequate water and light, nutrient deficiency can limit growth and crop yield.

Farmer spreading decomposing manure to improve soil fertility and plant nutrition

Processes

Plants take up essential elements from the soil through their roots and from the air (mainly consisting of nitrogen and oxygen) through their leaves. Nutrient uptake in the soil is achieved by cation exchange, wherein root hairs pump hydrogen ions (H^+) into the soil through proton pumps. These hydrogen ions displace cations attached to negatively charged soil particles so that the cations are available for uptake by the root. In the leaves, stomata open to take in carbon dioxide and expel oxygen. The carbon dioxide molecules are used as the carbon source in photosynthesis.

The root, especially the root hair, is the essential organ for the uptake of nutrients. The structure and architecture of the root can alter the rate of nutrient uptake. Nutrient ions are transported to the center of the root, the stele in order for the nutrients to reach the conducting tissues, xylem and phloem. The Casparian strip, a cell wall outside the stele but within the root, prevents passive flow of water and nutrients, helping to regulate the uptake of nutrients and water. Xylem moves water and inorganic molecules within the plant and phloem accounts for organic molecule transportation. Water potential plays a key role in a plant's nutrient uptake. If the water potential is more negative within the plant than the surrounding soils, the nutrients will move from the region of higher solute concentration—in the soil—to the area of lower solute concentration - in the plant.

There are three fundamental ways plants uptake nutrients through the root:

- Simple diffusion occurs when a nonpolar molecule, such as O_2, CO_2, and NH_3 follows a concentration gradient, moving passively through the cell lipid bilayer membrane without the use of transport proteins.
- Facilitated diffusion is the rapid movement of solutes or ions following a

concentration gradient, facilitated by transport proteins.

- Active transport is the uptake by cells of ions or molecules against a concentration gradient; this requires an energy source, usually ATP, to power molecular pumps that move the ions or molecules through the membrane.

Nutrients can be moved within plants to where they are most needed. For example, a plant will try to supply more nutrients to its younger leaves than to its older ones. When nutrients are mobile within the plant, symptoms of any deficiency become apparent first on the older leaves. However, not all nutrients are equally mobile. Nitrogen, phosphorus, and potassium are mobile nutrients while the others have varying degrees of mobility. When a less-mobile nutrient is deficient, the younger leaves suffer because the nutrient does not move up to them but stays in the older leaves. This phenomenon is helpful in determining which nutrients a plant may be lacking.

Many plants engage in symbiosis with microorganisms. Two important types of these relationship are

- with bacteria such as rhizobia, that carry out biological nitrogen fixation, in which atmospheric nitrogen (N_2) is converted into ammonium (NH^{+4}); and
- with mycorrhizal fungi, which through their association with the plant roots help to create a larger effective root surface area. Both of these mutualistic relationships enhance nutrient uptake.

Though nitrogen is plentiful in the Earth's atmosphere, relatively few plants harbour nitrogen-fixing bacteria, so most plants rely on nitrogen compounds present in the soil to support their growth. These can be supplied by mineralization of soil organic matter or added plant residues, nitrogen fixing bacteria, animal waste, through the breaking of triple bonded Nitrogen molecules by lightening strikes or through the application of fertilizers.

2Functions of Nutrients

At least 17 elements are known to be essential nutrients for plants. In relatively large amounts, the soil supplies nitrogen, phosphorus, potassium, calcium, magnesium, and sulfur; these are often called the macronutrients. In relatively small amounts, the soil supplies iron, manganese, boron, molybdenum, copper, zinc, chlorine, and cobalt, the so-called micronutrients. Nutrients must be available not only in sufficient amounts but also in appropriate ratios.

Plant nutrition is a difficult subject to understand completely, partially because of the variation between different plants and even between different species or individuals of a given clone. Elements present at low levels may cause deficiency symptoms, and toxicity is possible at levels that are too high. Furthermore, deficiency of one element may present as symptoms of toxicity from another element, and vice versa. An abundance of one nutrient may cause a deficiency of another nutrient. For example, K^+ uptake can be influenced by the amount of NH+ 4 available.

Although nitrogen is plentiful in the Earth's atmosphere, relatively few plants engage in nitrogen fixation (conversion of atmospheric nitrogen to a biologically useful form). Most plants, therefore, require nitrogen compounds to be present in the soil in which they grow.

Carbon and oxygen are absorbed from the air while other nutrients are absorbed from the soil. Green plants obtain their carbohydrate supply from the carbon dioxide in the air by the process of photosynthesis. Each of these nutrients is used in a different place for a different essential function.

Macronutrients (Derived from Air and Water)

Carbon

Carbon forms the backbone of most plant biomolecules, including proteins, starches and cellulose. Carbon is fixed through photosynthesis; this converts carbon dioxide from the air into carbohydrates which are used to store and transport energy within the plant.

Hydrogen

Hydrogen also is necessary for building sugars and building the plant. It is obtained almost entirely from water. Hydrogen ions are imperative for a proton gradient to help drive the electron transport chain in photosynthesis and for respiration.

Oxygen

Oxygen is a component of many organic and inorganic molecules within the plant, and is acquired in many forms. These include: O_2 and CO_2 (mainly from the air via leaves) and H_2O, NO_3^-, $H_2PO_4^-$ and SO_4^{2-} (mainly from the soil water via roots). Plants produce oxygen gas (O_2) along with glucose during photosynthesis but then require O_2 to undergo aerobic cellular respiration and break down this glucose to produce ATP.

Macronutrients (primary)

Nitrogen is a major constituent of several of the most important plant substances. For example, nitrogen compounds comprise 40% to 50% of the dry matter of protoplasm, and it is a constituent of amino acids, the building blocks of proteins. It is also an essential constituent of chlorophyll. Nitrogen deficiency most often results in stunted growth, slow growth, and chlorosis. Nitrogen deficient plants will also exhibit a purple appearance on the stems, petioles and underside of leaves from an accumulation of anthocyanin pigments. Most of the nitrogen taken up by plants is from the soil in the forms of NO^{-3}, although in acid environments such as boreal forests where nitrification is less likely to occur, ammonium NH^+ 4 is more likely to be the dominating source of nitrogen. Amino acids and proteins can only be built from NH^+ 4, so NO^- 3 must be reduced. In many agricultural settings, nitrogen is the limiting nutrient for rapid growth. Nitrogen is transported via the xylem from the roots to the leaf canopy as nitrate ions, or in an organic form, such as amino acids or amides. Nitrogen can also be transported in the phloem sap as amides, amino acids and ureides; it is therefore mobile within the plant, and the older leaves exhibit chlorosis and necrosis earlier than the younger leaves.

There is an abundant supply of nitrogen in the earth's atmosphere — N_2 gas comprises nearly 79% of air. However, N_2 is unavailable for use by most organisms because there is a triple bond between the two nitrogen atoms, making the molecule almost inert. In order for nitrogen to be used for growth it must be "fixed" (combined) in the form of ammonium (NH_4) or nitrate (NO_3) ions. The weathering of rocks releases these ions so slowly that it has a negligible effect on the availability of fixed nitrogen. Therefore, nitrogen is often the limiting factor for growth and biomass production in all environments where there is a suitable climate and availability of water to support life.

Nitrogen enters the plant largely through the roots. A "pool" of soluble nitrogen accumulates. Its composition within a species varies widely depending on several factors, including day length, time of day, night temperatures, nutrient deficiencies, and nutrient imbalance. Short day length promotes asparagine formation, whereas glutamine is produced under long day regimes. Darkness favors protein breakdown accompanied by high asparagine accumulation. Night temperature modifies the effects due to night length, and soluble nitrogen tends to accumulate owing to retarded synthesis and breakdown of proteins. Low night temperature conserves glutamine; high night temperature increases accumulation of asparagine because of breakdown. Deficiency of K accentuates differences between long- and short-day plants. The pool of soluble nitrogen is much smaller than in well-nourished plants when N and P are deficient since uptake of nitrate and further reduction and conversion of N to organic forms is restricted more than is protein synthesis. Deficiencies of Ca, K, and S affect the conversion of organic N to protein more than uptake and reduction. The size of the pool of soluble N is no guide *per se* to growth rate, but the size of the pool in relation to total N might be a useful ratio in this regard. Nitrogen availability in the rooting medium also affects the size and structure of tracheids formed in the long lateral roots of white spruce (Krasowski and Owens 1999).

Microorganisms have a central role in almost all aspects of nitrogen availability, and therefore for life support on earth. Some bacteria can convert N_2 into ammonia by the process termed *nitrogen fixation*; these bacteria are either free-living or form symbiotic associations with plants or other organisms (e.g., termites, protozoa), while other bacteria bring about transformations of ammonia to nitrate, and of nitrate to N_2 or other nitrogen gases. Many bacteria and fungi degrade organic matter, releasing fixed nitrogen for reuse by other organisms. All these processes contribute to the nitrogen cycle.

Phosphorus

Like nitrogen, phosphorus is involved with many vital plant processes. Within a plant, it is present mainly as a structural component of the nucleic acids: deoxyribonucleic acid (DNA) and ribonucleic acid (RNA), as well as a constituent of fatty phospholipids, that are important in membrane development and function. It is present in both organic and inorganic forms, both of which are readily translocated within the plant. All energy transfers in the cell are critically dependent on phosphorus. As with all living things, phosphorus is part of the Adenosine triphosphate (ATP), which is of immediate use in all processes that require energy with the cells. Phosphorus can also be used to modify the activity of various enzymes by phosphorylation, and is used for cell signaling. Phosphorus is concentrated at the most actively growing points of a plant and stored within seeds in anticipation of their germination. Phosphorus is most commonly found in the soil in the form of polyprotic phosphoric acid (H_3PO_4), but is taken up most readily in the form of H_2PO^{-4}. Phosphorus is available to plants in limited quantities in most soils because it is released very slowly from insoluble phosphates and is rapidly fixed once again. Under most environmental conditions it is the element that limits growth because of this constriction and due to its high demand by plants and microorganisms. Plants can increase phosphorus uptake by a mutualism with mycorrhiza. A Phosphorus deficiency in plants is characterized by an intense green coloration or reddening in leaves due to lack of chlorophyll. If the plant is experiencing high phosphorus deficiencies the leaves may become denatured and show signs of death. Occasionally the leaves may appear purple from an accumulation of anthocyanin. Because phosphorus is a mobile nutrient, older leaves will show the first signs of deficiency.

On some soils, the phosphorus nutrition of some conifers, including the spruces, depends on the ability of mycorrhizae to take up, and make soil phosphorus available to the tree, hitherto unobtainable to the non-mycorrhizal root. Seedling white spruce, greenhouse-grown in sand testing negative for phosphorus, were very small and purple for many months until spontaneous mycorrhizal inoculation, the effect of which was manifested by a greening of foliage and the development of vigorous shoot growth.

Phosphorus deficiency can produce symptoms similar to those of nitrogen deficiency, but as noted by Russel:"Phosphate deficiency differs from nitrogen deficiency in being extremely difficult to diagnose, and crops can be suffering from extreme starvation without there being any obvious signs that lack of phosphate is the cause". Russell's observation applies to at least some coniferous seedlings, but Benzian found that although response to phosphorus in very acid forest tree nurseries in England was consistently high, no species (including Sitka spruce) showed any visible symptom of deficiency other than a slight lack of lustre. Phosphorus levels have to be exceedingly low before visible symptoms appear in such seedlings. In sand culture at 0 ppm phosphorus, white spruce seedlings were very small and tinted deep purple; at 0.62 ppm, only the smallest seedlings were deep purple; at 6.2 ppm, the seedlings were of good size and color.

It is useful to apply a high phosphorus content fertilizer, such as bone meal, to perennials to help with successful root formation.

Potassium

Unlike other major elements, potassium does not enter into the composition of any of the important plant constituents involved in metabolism, but it does occur in all parts of plants in substantial amounts. It seems to be of particular importance in leaves and at growing points. Potassium is outstanding among the nutrient elements for its mobility and solubility within plant tissues. Processes involving potassium include the formation of carbohydrates and proteins, the regulation of internal plant moisture, as a catalyst and condensing agent of complex substances, as an accelerator of enzyme action, and as contributor to photosynthesis, especially under low light intensity.

Potassium regulates the opening and closing of the stomata by a potassium ion pump. Since stomata are important in water regulation, potassium regulates water loss from the leaves and increases drought tolerance. Potassium deficiency may cause necrosis or interveinal chlorosis. The potassium ion (K^+) is highly mobile and can aid in balancing the anion (negative) charges within the plant. Potassium helps in fruit coloration, shape and also increases its brix. Hence, quality fruits are produced in potassium-rich soils. Potassium serves as an activator of enzymes used in photosynthesis and respiration. Potassium is used to build cellulose and aids in photosynthesis by the formation of a chlorophyll precursor. Potassium deficiency may result in higher risk of pathogens, wilting, chlorosis, brown spotting, and higher chances of damage from frost and heat. When soil-potassium levels are high, plants take up more potassium than needed for healthy growth. The term *luxury consumption* has been applied to this. When potassium is moderately deficient, the effects first appear in the older tissues, and from there progress towards the growing points. Acute deficiency severely affects growing points, and die-back commonly occurs. Symptoms of potassium deficiency in white spruce include: browning and death of needles (chlorosis); reduced growth in height and diameter; impaired retention of needles; and reduced needle length. A relationship between potassium nutrition and cold resistance has been found in several tree species, including two species of spruce.

Macronutrients (Secondary and Tertiary)

Sulphur

Sulphur is a structural component of some amino acids and vitamins, and is essential in the manufacturing of chloroplasts. Sulphur is also found in the iron-sulphur complexes of the electron transport chains in photosynthesis. Sulphate ions are mobile and its deficiency, therefore, affects older tissues first. Symptoms of deficiency include yellowing of leaves and stunted growth.

Calcium

Calcium regulates transport of other nutrients into the plant and is also involved in the activation of certain plant enzymes. Calcium deficiency results in stunting. This nutrient is involved in photosynthesis and plant structure. Blossom end rot is also a result of inadequate calcium.

Calcium in plants occurs chiefly in the leaves, with lower concentrations in seeds, fruits, and roots. A major function is as a constituent of cell walls. When coupled with certain acidic compounds of the jelly-like pectins of the middle lamella, calcium forms an insoluble salt. It is also intimately involved in meristems, and is particularly important in root development, with roles in cell division, cell elongation, and the detoxification of hydrogen ions. Other functions attributed to calcium are; the neutralization of organic acids; inhibition of some potassium-activated ions; and a role in nitrogen absorption. A notable feature of calcium-deficient plants is a defective root system. Roots are usually affected before above-ground parts.

Magnesium

The outstanding role of magnesium in plant nutrition is as a constituent of the chlorophyll molecule. As a carrier, it is also involved in numerous enzyme reactions as an effective activator, in which it is closely associated with energy-supplying phosphorus compounds. Magnesium is very mobile in plants, and, like potassium, when deficient is translocated from older to younger tissues, so that signs of deficiency appear first on the oldest tissues and then spread progressively to younger tissues.

Micro-nutrients

Plants are able sufficiently to accumulate most trace elements. Some plants are sensitive indicators of the chemical environment in which they grow (Dunn 1991), and some plants have barrier mechanisms that exclude or limit the uptake of a particular element or ion species, e.g., alder twigs commonly accumulate molybdenum but not arsenic, whereas the reverse is true of spruce bark (Dunn 1991). Otherwise, a plant can integrate the geochemical signature of the soil mass permeated by its root system together with the contained groundwaters. Sampling is facilitated by the tendency of many elements to accumulate in tissues at the plant's extremities.

Iron

Iron is necessary for photosynthesis and is present as an enzyme cofactor in plants. Iron deficiency can result in interveinal chlorosis and necrosis. Iron is not a structural part of chlorophyll but very much essential for its synthesis. Copper deficiency can be responsible for promoting an iron deficiency.

Molybdenum

Molybdenum is a cofactor to enzymes important in building amino acids and is involved in nitrogen metabolism. Molybdenum is part of the nitrate reductase enzyme (needed for the reduction of nitrate) and the nitrogenase enzyme (required for biological nitrogen fixation).

Boron

Boron is found in the highly insoluble mineral, tourmaline. It is absorbed by plants in the form of the anion BO3 3. It is available to plants in moderately soluble mineral forms of Ca, Mg and Na borates and the highly soluble form of organic compounds. Concentration in soil must, in general, be below 5 ppm in the soil water solution, above that toxicity results. Its availability in soils ranges from 20 to 200 pounds per acre in the first eight inches, of which less than 5% is available. It is available to plants over a range of pH, from 5.0 to 7.5. It is mobile in the soil, hence, it is prone to leaching. Leaching removes substantial amounts of boron in sandy soil, but little in fine silt or clay soil. Boron's fixation to those minerals at high pH can render boron unavailable, while low pH frees the fixed boron, leaving it prone to leaching in wet climates. It precipitates with other minerals in the form of borax in which form it was first used over 400 years ago as a soil supplement. Decomposition of organic material causes boron to be deposited in the topmost soil layer; organic forms of boron are more soluble than their mineral form, hence are more available in the top few inches. When soil dries it can cause a precipitous drop in the availability of boron to plants as the plants cannot draw nutrients from that desiccated layer. Hence, boron deficiency diseases appear in dry weather.

Boron has many functions within a plant: it affects flowering and fruiting, pollen germination, cell division, and active salt absorption. The metabolism of amino acids and proteins, carbohydrates, calcium, and water are strongly affected by boron. Many of those listed functions may be embodied by its function in moving the highly polar sugars through cell membranes by reducing their polarity and hence the energy needed to pass the sugar. If sugar cannot pass to the fastest growing parts rapidly enough, those parts die. Boron is relatively immobile within a plant suggesting that the molecule is fixed to the points in the membrane where they facilitate sugar transport.

Boron is not relocatable in the plant via the phloem. It must be supplied to the growing parts via the xylem. Foliar sprays affect only those parts sprayed, which may be insufficient for the fastest growing parts, and is very temporary.

Boron is essential for the proper forming and strengthening of cell walls. Lack of boron results in short thick cells producing stunted fruiting bodies and roots. Calcium to boron ratio must be maintained in a narrow range for normal plant growth. For alfalfa, that calcium to boron ratio must be from 80:1 to 600:1. Boron deficiency appears at 800:1 and higher. For alfalfa, similar ratios exist for magnesium, copper, nitrogen and potassium. Boron levels within plants differ with plant species and range from 2.3 p.p.m for barley to 94.7 p.p.m for poppy . Lack of boron causes failure of calcium metabolism which produces hollow heart in beets and peanuts.

Inadequate amounts of boron affect many agricultural crops, legume forage crops most strongly. Of the micronutrients, boron deficiencies are second most common after zinc. Deficiencies of boron when soil is cropped are common and require the application of mineral supplement; one ton of alfalfa hay carries with it one ounce of boron, 100 bushels of peaches 4 ounces. Deficiency results in the death of the terminal growing points. Symptoms first appear as stunted growth, then to cellular changes, which leads to physical changes, and finally death of the plant.

Boron supplements derive from dry lake bed deposits such as those in Death Valley, USA, in the form of sodium tetraborate (borax), from which less soluble calcium borate is made. Foliar sprays are used on fruit crop trees in soils of high alkalinity. Boron is often applied to fields as a contaminant in other soil amendments but is not generally adequate to make up the rate of loss by cropping. The rates of application of borate to produce an adequate alfalfa crop range from 15 pounds per acre for a sandy-silt, acidic soil of low organic matter, to 60 pounds per acre for a soil with high organic matter, high cation exchange capacity and high pH.

Boron concentration in soil water solution higher than one ppm is toxic to most plants. Toxic concentrations within plants are 10 to 50 ppm for small grains and 200 ppm in boron-tolerant crops such as sugar beets, rutabaga, cucumbers, and conifers. Toxic soil conditions are generally limited to arid regions or can be caused by underground borax deposits in contact with water or volcanic gases dissolved in percolating water. Application rates should be limited to a few pounds per acre in a test plot to determine if boron is needed generally. Otherwise, testing for boron levels in plant material is required to determine remedies. Excess boron can be removed by irrigation and assisted by application of elemental sulfur to lower the pH and increase boron's solubility. Application of calcium will increase soil alkalinity, causing boron to fix on the mineral soil component and remove some fraction, thereby reducing boron toxicity.

Boron deficiencies must be detected by analysis of plant material to apply a correction before the obvious symptoms appear, after which it is too late to prevent crop loss. Strawberries deficient in boron will produce lumpy fruit; apricots will not blossom or, if they do, will not fruit or will drop their fruit depending on the level of boron deficit. Broadcast of boron supplements is effective and long term; a foliar spray is immediate but must be repeated.

Boron is an essential element for the health of animals which derive their boron from plant material.

Copper

Copper is important for photosynthesis. Symptoms for copper deficiency include chlorosis. It is involved in many enzyme processes; necessary for proper photosynthesis; involved in the manufacture of lignin (cell walls) and involved in grain production. It is also hard to find in some soil conditions.

Manganese

Manganese is necessary for photosynthesis, including the building of chloroplasts. Manganese deficiency may result in coloration abnormalities, such as discolored spots on the foliage.

Sodium

Sodium is involved in the regeneration of phosphoenolpyruvate in CAM and C4 plants. Sodium can potentially replace potassium's regulation of stomatal opening and closing.

Essentiality of sodium:

- Essential for C4 plants rather C3
- Substitution of K by Na: Plants can be classified into four groups:

1. Group A—a high proportion of K can be replaced by Na and stimulate the growth, which cannot be achieved by the application of K
2. Group B—specific growth responses to Na are observed but they are much less distinct
3. Group C—Only minor substitution is possible and Na has no effect
4. Group D—No substitution occurs

- Stimulate the growth—increase leaf area and stomata. Improves the water balance
- Na functions in metabolism

1. C4 metabolism
2. Impair the conversion of pyruvate to phosphoenol-pyruvate
3. Reduce the photosystem II activity and ultrastructural changes in mesophyll chloroplast

- Replacing K functions

1. Internal osmoticum

2. Stomatal function
3. Photosynthesis
4. Counteraction in long distance transport
5. Enzyme activation

- Improves the crop quality e.g. improves the taste of carrots by increasing sucrose

Zinc

Zinc is required in a large number of enzymes and plays an essential role in DNA transcription. A typical symptom of zinc deficiency is the stunted growth of leaves, commonly known as "little leaf" and is caused by the oxidative degradation of the growth hormone auxin.

Nickel

In higher plants, nickel is absorbed by plants in the form of Ni^{2+} ion. Nickel is essential for activation of urease, an enzyme involved with nitrogen metabolism that is required to process urea. Without nickel, toxic levels of urea accumulate, leading to the formation of necrotic lesions. In lower plants, nickel activates several enzymes involved in a variety of processes, and can substitute for zinc and iron as a cofactor in some enzymes.

Chlorine

Chlorine, as compounded chloride, is necessary for osmosis and ionic balance; it also plays a role in photosynthesis.

Cobalt

Cobalt has proven to be beneficial to at least some plants although it does not appear to be essential for most species. It has, however, been shown to be essential for nitrogen fixation by the nitrogen-fixing bacteria associated with legumes and other plants.

Aluminium

- Tea has a high tolerance for aluminum (Al) toxicity and the growth is stimulated by Al application. The possible reason is the prevention of Cu, Mn or P toxicity effects.
- There have been reports that Al may serve as a fungicide against certain types of root rot.

Silicon

Silicon is not considered an essential element for plant growth and development. It is always found in abundance in the environment and hence if needed it is available. It is found in the structures of plants and improves the health of plants. In plants, silicon has been shown in experiments to strengthen cell walls, improve plant strength, health, and productivity.There have been studies showing evidence of silicon improving drought and frost resistance, decreasing lodging potential and boosting the plant's natural pest and disease fighting systems. Silicon has also been shown to improve plant vigor and physiology by improving root mass and density, and increasing above ground plant biomass and crop yields. Silicon is currently under consideration by the Association of American Plant Food Control Officials (AAPFCO) for elevation to the status of a "plant beneficial substance".

Vanadium

Vanadium may be required by some plants, but at very low concentrations. It may also be substituting for molybdenum.

Selenium

Selenium is probably not essential for flowering plants, but it can be beneficial; it can stimulate plant growth, improve tolerance of oxidative stress, and increase resistance to pathogens and herbivory.

Selenium is, however, an essential mineral element for animal (including human) nutrition and selenium deficiencies are known to occur when food or animal feed is grown on selenium-deficient soils. The use of inorganic selenium fertilizers can increase selenium concentrations in edible crops and animal diets thereby improving animal health.

Nutrient Deficiency

The effect of a nutrient deficiency can vary from a subtle depression of growth rate to obvious stunting, deformity, discoloration, distress, and even death. Visual symptoms distinctive enough to be useful in identifying a deficiency are rare. Most deficiencies are multiple and moderate. However, while a deficiency is seldom that of a single nutrient, nitrogen is commonly the nutrient in shortest supply.

Chlorosis of foliage is not always due to mineral nutrient deficiency. Solarization can produce superficially similar effects, though mineral deficiency tends to cause premature defoliation, whereas solarization does not, nor does solarization depress nitrogen concentration.

Nutrient Status of Plants

Nutrient status (mineral nutrient and trace element composition, also called ionome and nutrient profile) of plants are commonly portrayed by tissue elementary analysis. Interpretation of the results of such studies, however, has been controversial. During the last decades the nearly two-century-old "law of minimum" or "Liebig's law" (that states that plant growth is controlled not by the total amount of resources available, but by the scarcest resource) has been replaced by several mathematical approaches that use different models in order to take the interactions between the individual nutrients into account. The latest developments in this field are based on the fact that the nutrient elements (and compounds) do not act independently from each other; Baxter, 2015, because there may be direct chemical interactions between them or they may influence each other's uptake, translocation, and biological action via a number of mechanisms as exemplified for the case of ammonia.

Plant Nutrition in Agricultural Systems

Hydroponics

Hydroponics is a method for growing plants in a water-nutrient solution without the use of nutrient-rich soil. It allows researchers and home gardeners to grow their plants in a controlled environment. The most common solution is the Hoagland solution, developed by D. R. Hoagland in 1933. The solution consists of all the essential nutrients in the correct proportions necessary for most plant growth. An aerator is used to prevent an anoxic event or hypoxia. Hypoxia can affect nutrient uptake of a plant because, without oxygen present, respiration becomes inhibited within the root cells. The nutrient film technique is a hydroponic technique in which the roots are not fully submerged. This allows for adequate aeration of the roots, while a "film" thin layer of nutrient-rich water is pumped through the system to provide nutrients and water to the plant.

REFERENCES

1. Frank B. Salisbury; Cleon W. Ross (1992). Plant physiology. Brooks/Cole Pub Co. ISBN 0-534-15162-0.
2. Trevor Robinson (1963). The organic constituents of higher plants: their chemistry and interrelationships. Cordus Press. p. 183.
3. Kimler, L. M. (1975). "Betanin, the red beet pigment, as an antifungal agent". Botanical Society of America, Abstracts of papers. 36.
4. Fosket, Donald E. (1994). Plant Growth and Development: A Molecular Approach. San Diego: Academic Press. pp. 498–509. ISBN 0-12-262430-0.
5. plantphys.net
6. Adrian Charles Slack; Jane Gate (1980). Carnivorous Plants. Cambridge, Mass. : MIT Press. p. 160. ISBN 978-0-262-19186-9.
7. Kingsley Rowland Stern; Shelley Jansky (1991). Introductory Plant Biology. WCB/McGraw-Hill. p. 309. ISBN 978-0-697-09948-8.
8. Hales, Stephen. 1727. Vegetable Staticks http://www.illustratedgarden.org/mobot/rarebooks/title.asp?relation=QK711H341727
9. Duane Isely (1994). 101 Botanists. Iowa State Press. pp. 216–219. ISBN 978-0-8138-2498-7.

CHAPTER 6

Plant Taxonomy

Chapter Contents

- **Introduction**
- **Identification, Classification and Description**
- **List of Systems of Plant Taxonomy**
- **Chronological List Of Systems**

Taxonomy is about grouping things; plant taxonomy is about grouping plants. The primary goal of a plant taxonomist is to try and summarize the variation in the plant world and express it in a manner that reflects the phylogenetic relationships among the various taxa observed. Let us look at the last sentence in more detail.

INTRODUCTION

Plant taxonomy is the science that finds, identifies, describes, classifies, and names plants. Thus making it one of the main branches of taxonomy(the science that finds, describes, classifies, and names living things).

Plant taxonomy is closely allied to plant systematics, and there is no sharp boundary between the two. In practice, "Plant systematics" involves relationships between plants and their evolution, especially at the higher levels, whereas "plant taxonomy" deals with the actual handling of plant specimens. The precise relationship between taxonomy and systematics, however, has changed along with the goals and methods employed. Plant taxonomy is well known for being turbulent, and traditionally not having any close agreement on circumscription and placement of taxa. See the list of systems of plant taxonomy.

IDENTIFICATION, CLASSIFICATION AND DESCRIPTION

Three goals of plant taxonomy are the identification, classification and description of plants. The distinction between these three goals is important and often overlooked.

Plant identification is the determination of the identity of an unknown plant by comparison with previously collected specimens or with the aid of books or identification manuals. The process of identification connects the specimen with a published name. Once a plant specimen has been identified, its name and properties are known.

Plant classification is the placing of known plants into groups or categories to show some relationship. Scientific classification follows a system of rules that standardizes the results, and groups successive categories into a hierarchy. For example, the family to which the lilies belong is classified as follows:

- Kingdom: Plantae
- Phylum: Magnoliophyta
- Class: Liliopsida
- Series: Liliales
- Family: Liliaceae
- Genus: ......
- Species: ......

The classification of plants results in an organized system for the naming and cataloging of future specimens, and ideally reflects scientific ideas about inter-relationships between plants. The set of rules and recommendations for formal botanical nomenclature, including plants, is governed by the International Code of Nomenclature for algae, fungi, and plants abbreviated as ICN.

Plant description is a formal description of a newly discovered species, usually in the form of a scientific paper using ICN guidelines. The names of these plants are then registered on the International Plant Names Index along with all other validly published names.

LIST OF SYSTEMS OF PLANT TAXONOMY

A pioneering system of plant taxonomy, Linnaeus's *Systema Naturae*, Leiden, 1735

This **list of systems of plant taxonomy** presents "taxonomic systems" used in plant classification.

A taxonomic system is a coherent whole of taxonomic judgments on circumscription and placement of the considered taxa. It is only a "system" if it is applied to a large group of such taxa (for example, all the flowering plants).

There are two main criteria for this list. A system must be taxonomic, that is deal with a large number of plants, by their botanical names. Secondly it must be a system, i.e. deal with the relationships of plants. Although thinking about relationships of plants had started much earlier (see history of plant systematics), such systems really only came into being in the 19th century, as a result of an ever increasing influx from all over the world of newly discovered plant species. The 18th century saw some early systems, which are perhaps precursors rather than full taxonomic systems.

A milestone event was the publication of *Species Plantarum* by Linnaeus which serves as the starting point of binomial nomenclature for plants. By its size this would qualify to be on this list, but it does not deal with relationships, beyond assigning plants into genera.

Note that a system is not necessarily monolithic and often goes through several stages of development, resulting in several versions of the same system. When a system is widely adopted, many authors will adopt their own particular version of the system. The Cronquist system is well known for existing in many versions.

CHRONOLOGICAL LIST OF SYSTEMS

Pre-Linnaean

Theophrastus classification

Historia Plantarum (*Enquiry into Plants*), c. 300 BC

Causes of Plants, c. 300 BC

Dioscorides classification

De Materia Medica, c. 60 AD

Albertus Magnus classification

De Vegetabilibus, c. 1256

Caesalpino classification

De Plantis, 1583

John Gerard classification

Herball, or Generall Historie of Plantes, 1597

Gaspard Bauhin classification

Pinax theatri botanici, 1623

John Ray classification

Historia Plantarum, 1686–1704

Tournefort classification

Éléments de botanique, 1694

From Linnaeus to Darwin (pre-Darwinian)

- Linnaean systems

Systema Naturae, 1st edition, 1735.

Systema Naturae, 10th edition, 1758 (vol. 1), 1759 (vol. 2). Starting point of zoological nomenclature.

Species Plantarum, 1753. Starting point of botanical nomenclature.

Genera Plantarum, 1737 (1st ed.), 1753 (5th ed.).

Philosophia Botanica, 1751.

- Adanson system

Familles naturelles des plantes, 1763.

- de Jussieu system

A.L. de Jussieu (1789). Genera Plantarum, secundum ordines naturales disposita juxta methodum in Horto Regio Parisiensi exaratam.

(available online at Gallica)

- de Candolle system

A. P. de Candolle (1819). Théorie élémentaire de la botanique, ou exposition des principes de la classification naturelle et de l'art de décrire et d'etudier les végétaux (2nd ed.).

A. P. de Candolle; et al. (1824–1873). Prodromus systemati naturalis regni vegetabilis sive enumeratio contracta ordinum, generum specierumque plantarum huc usque cognitarum, juxta methodi naturalis normas digesta.

(available online at Gallica)

- Berchtold and Presl system

Berchtold and Presl. O Prirozenosti Rostlin 1820

- Dumortier system

Dumortier, Barthélemy-Charles (1829). Analyse des familles des plantes :avec l'indication des principaux genres qui s'y rattachent (in French). Tournay: Casterman. Retrieved 16 January 2016.

- Lindley system

Lindley (1830). An Introduction to the Natural System of Botany.

(available online at BHL)

Lindley (1845). The Vegetable Kingdom.

(available online at BHL)

- Bentham & Hooker system

G. Bentham & J.D. Hooker (1862–1883). Genera plantarum ad exemplaria imprimis in herbariis kewensibus servata definita. 3 volumes.

(available online at Gallica)

- Baillon system

H. Baillon (1867–1894). Histoire des plantes. 13 volumes.

Post Darwinian (Phyletic)

- Eichler system

Eichler (1883). Syllabus der Vorlesungen über Phanerogamenkunde (3rd ed.).

- Van Tieghem system

Van Tieghem, Philippe Édouard Léon; (1884; 2nd ed., 1891). Traité de botanique (in French). Paris: Savy. Retrieved 27 December 2015.

- Engler system

A. Engler & K. Prantl; (1887–1915; 2nd edition, 1924–). Die Natürlichen Pflanzenfamilien.

A. Engler (1892). Syllabus der Vorlesungen über specielle und medicinisch-pharmaceutische Botanik [or, in further editions, Syllabus der Pflanzenfamilien].

- Dalla Torre & Harms system

K.W. von Dalla Torre & H. Harms (1900–1907). Genera Siphonogamarum, ad systema Englerianum conscripta.

- Bessey system

Charles E. Bessey (1907). "A Synopsis of Plant Phyla". Univ. Nebraska Studies 7: 275–358.

Charles E. Bessey (1915). "The phylogenetic taxonomy of flowering plants". Annals of the Missouri Botanical Garden. Missouri Botanical Garden Press. ***2*** *(1/2): 109–164. doi:10.2307/2990030. JSTOR 2990030.*

- Wettstein system

R. Wettstein; (in two volumes, 1901-1908; 2nd edition, 1911; 3rd edition, 1923-24; 4th edition, 1933-35). Handbuch der systematischen Botanik.

- Rendle system

Alfred Barton Rendle. *The Classification of Flowering Plants* 1904, 1925

- Lotsy system

Johannes Paulus Lotsy. *Vorträge über botanische Stammesgeschichte, gehalten an der Reichsuniversität zu Leiden. Ein Lehrbuch der Pflanzensystematik.* 1907-1911

- Hallier system

*Hallier, H (1912). "L'origine et le système phylétique des angiospermes exposés à l'aide de leur arbre généalogique". Archives Néerlandaises des Sciences Exactes et Naturelles, Série III. B. **1**: 146–234.*

- Warming system

Warming, Eugenius (1912). Frøplanterne (Spermatofyter) (in Danish). Kjøbenhavn: Gyldendalske Boghandel Nordisk Forlag. Retrieved 28 December 2015.

- Hutchinson system

J. Hutchinson; (two volumes, 1926–1934; 2nd edition 1959; 3rd edition, 1973). The families of flowering plants, arranged according to a new system based on their probable phylogeny.

- Calestani system

*Calestani, Vittorio (1933). «Le origini e la classificazione delle Angiosperme". Archo. Bot. Sist. Fito-Geogr. Genet. **9**: 274–311.*

- Kimura system

*Kimura, Yojiro (1956). «Système et phylogénie des monocotyledones". Notulae Systematicae, Herbier du Muséum de Paris. **15**: 137–159.*

- Benson system

Lyman David Benson. *Plant Classification* 1957

- Emberger system

Chaudefaud, Marius; Emberger, Louis (1960). Traité de Botanique systématique 2 vols. Paris: Masson.

- Melchior system

H. Melchior (1964). Adolf Engler, ed. Syllabus der Pflanzenfamilien (12th edition, II. Band ed.). (also known as modified Engler system, in Angiospermae)

- Takhtajan system

A. Takhtajan; (in Russian, 1966). A system and phylogeny of the flowering plants.

A. Takhtajan; (1969, transl. from Russian by C. Jeffrey). Flowering plants: origin and dispersal. ISBN 0-05-001715-2.

A. Takhtajan (1980). "Outline of the classification of flowering plants (Magnoliophya)". Botanical

Review. ***46*** *(3): 225–359. doi:10.1007/BF02861558.*

A. Takhtajan (1997). Diversity and classification of flowering plants. ISBN 0-231-10098-1.

- Cronquist system

A. Cronquist; (1968; 2nd edition 1988). The evolution and classification of flowering plants. ISBN 0-89327-332-5.

A. Cronquist (1981). An integrated system of classification of flowering plants. ISBN 0-231-03880-1.

- Thorne system

R.F. Thorne (1968). "Synopsis of a putative phylogenetic classification of flowering plants". Aliso. ***6*** *(4): 57–66.*

R.F. Thorne (1983). "Proposed new alignments in the angiosperms". Nordic J. Bot. ***3****: 85–117. doi:10.1111/j.1756-1051.1983.tb01447.x.*

R.F. Thorne (1992). "Classification and geography of flowering plants". Botanical Review. ***58*** *(3): 225–348. doi:10.1007/BF02858611.*

R.F. Thorne (1992). "An updated phylogenetic classification of the flowering plants". Aliso. ***13****: 365–389.*

R.F. Thorne (2000). "The classification and geography of the flowering plants: dicotyledons of the class Angiospermae". Botanical Review. ***66*** *(4): 441–647. doi:10.1007/BF02869011.*

- Stebbins system

Stebbins, G.L. (1974). *Flowering plants: evolution above the species level.* Cambridge, Massachusetts: Harvard University Press, . [System followed by Heywood, V.H. (ed., 1978). *Flowering plants of the world.* Oxford: Oxford University Press, .]

- Dahlgren system

R.M.T. Dahlgren (1975). "A system of classification of angiosperms to be used to demonstrate the distribution of characters". Bot. Notiser. ***128****: 119–147.*

R.M.T. Dahlgren (1980). "A revised system of classification of angiosperms". Bot. J. Linn. Soc. ***80*** *(2): 91–124. doi:10.1111/j.1095-8339.1980.tb01661.x.*

R.M.T. Dahlgren (1983). "General Aspects of Angiosperm Evolution and Macrosystematics". Nordic Journal of Botany. ***3****: 119–149. doi:10.1111/j.1756-1051.1983.tb01448.x.*

R.M.T. Dahlgren & al. (1985). The families of the monocotyledons: structure, evolution, and taxonomy. ISBN 0-387-13655-X.

G. Dahlgren (1989). "An updated Angiosperm Classification". Botanical Journal of the Linnean Society. ***100*** *(3): 197–203. doi:10.1111/j.1095-8339.1989.tb01717.x.*

- Goldberg system

Aaron Goldberg (1986). "Classification, Evolution and Phylogeny of the Families of Dicotyledons". Smithsonian Contributions to Botany. ***58****: 1–314. doi:10.5479/si.0081024x.58.* (available online: Full text (PDF) here) [there is also a comparison among 11 Dicotyledons systems since 1960 until 1985]

Aaron Goldberg (1989). "Classification, Evolution and Phylogeny of the Families of Monocotyledons". Smithsonian Contributions to Botany. ***71****: 1–73. doi:10.5479/si.0081024x.71.* (available online: Full text (PDF) here)

- Kubitzki system (1990-)

K. Kubitzki & al.; (1990 ongoing). The Families and Genera of Vascular Plants.

- Shipunov system (1991-)
- Reveal system (1999)

Reveal System of Classification 1999

- Judd system (1999-2008)

(Modified APG)

Judd, Walter S. (2008). Plant systematics: a phylogenetic approach. 3rd ed. Sinauer Associates. ISBN 0-87893-407-3. Retrieved 29 January 2014.

- APG system

APG I (1998)

APG (1998). "An ordinal classification for the families of flowering plants". Annals of the Missouri Botanical Garden. Missouri Botanical Garden Press. ***85*** *(4): 531–553. doi:10.2307/2992015. JSTOR 2992015.*

W.S. Judd & al.; (1999; 2nd edition, 2002). Plant systematics: a phylogenetic approach. ISBN 0-87893-403-0.

APG II (2003)

APG (2003). "An update of the Angiosperm Phylogeny Group classification for the orders and families of flowering plants: APG II". Botanical Journal of the Linnean Society. ***141*** *(4): 399–436. doi:10.1046/j.1095-8339.2003.t01-1-00158.x.*

APG III (2009)

APG (2009). "An update of the Angiosperm Phylogeny Group classification for the orders and families of flowering plants: APG III.". Botanical Journal of the Linnean Society. ***161*** *(2): 399–436. doi:10.1111/j.1095-8339.2009.00996.x.*

(Available online:)

APG IV (2016)

APG IV (2016). "An update of the Angiosperm Phylogeny Group classification for the orders and families of flowering plants: APG IV". Botanical Journal of the Linnean Society. ***181*** *(1): 1–20. doi:10.1111/boj.12385. Retrieved 13 June 2016.*

REFERENCES

1. Dahlgren, Rolf; Clifford, H. T. (1982). The monocotyledons: A comparative study. London and New York: Academic Press. ISBN 9780122006807.
2. Reddy, S.M. (2007), University botany: Angiosperms, plant embryology and plant physiology, New Age International, p. 7, ISBN 978-81-224-1547-6
3. Stafleu, Frans A.; Cowan, Richard S. (1976–1988). Taxonomic literature: a selective guide to botanical publications and collections with dates, commentaries and types: Taxon. Lit., ed. 2 (TL2) (2nd ed.). Utrecht: Bohn, Scheltema & Holkema. ISBN 90 313 0224 4.

CHAPTER 7

Plant Anatomy

Chapter Contents

- **Inroduction**
- **Structural Divisions**
- **History**

The science of the structure of the organized plant body learned by dissection is called Plant Anatomy (anatomy-dissection). In general, Plant Anatomy refers to study of internal morphology, pertaining to different tissues. The subject of this chapter is structure of Angiosperms, with emphasis on primary tissues.

INRODUCTION

Plant anatomy or phytotomy is the general term for the study of the internal structure of plants. While originally it included plant morphology, which is the description of the physical form and external structure of plants, since the mid-20th century the investigations of plant anatomy are considered a separate, distinct field, and plant anatomy refers to *just* the internal plant structures. Plant anatomy is now frequently investigated at the cellular level, and often involves the sectioning of tissues and microscopy.

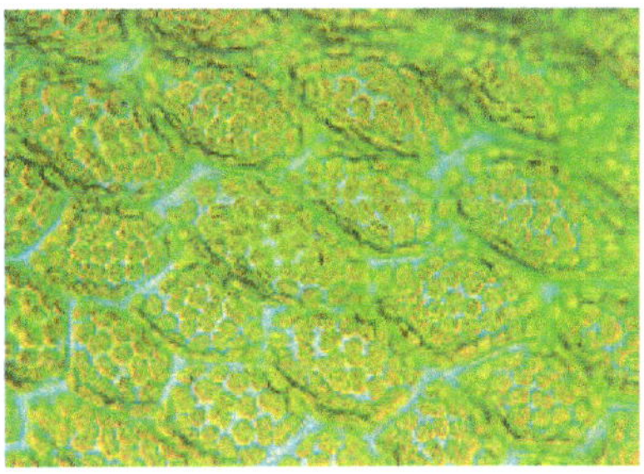

Chloroplasts in leaf cells of the moss *Mnium stellare*

STRUCTURAL DIVISIONS

Plant anatomy is often divided into the following categories:

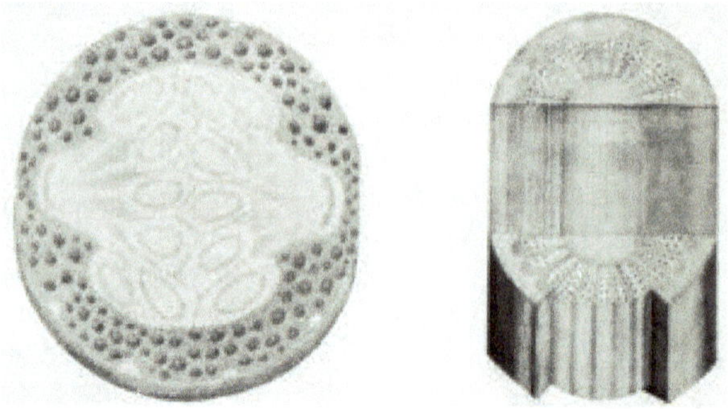

Vascular tissue of a gooseberry vine from Grew's*Anatomy of Plants*

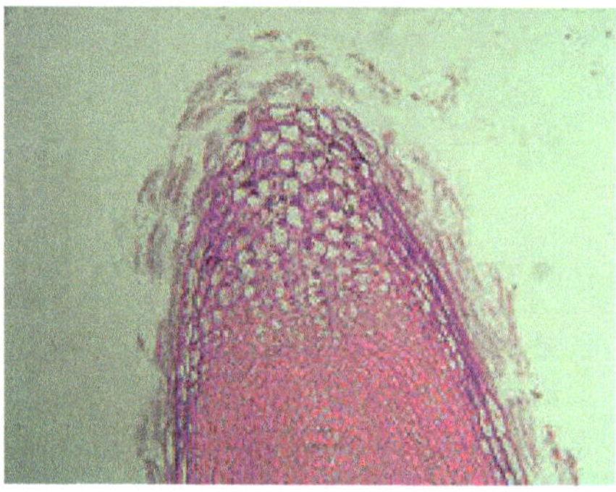

Root tip

- Flower anatomy
- Calyx
- Corolla
- Androecium
- Gynoecium
- Leaf anatomy
- Stem anatomy
- Stem structure
- Fruit/Seed anatomy
- Ovule
- Seed structure

Pericarp

Accessory fruit

Wood anatomy

Bark

Cork

Phloem

Vascular cambium

Heartwood and sapwood

branch collar

Root anatomy

Root structure

HISTORY

About 300 BC Theophrastus wrote a number of plant treatises, only two of which survive, *Enquiry into Plants*. He developed concepts of plant morphology and classification, which did not withstand the scientific scrutiny of the Renaissance.

A Swiss physician and botanist, Gaspard Bauhin, introduced binomial nomenclature into plant taxonomy. He published *Pinax theatri botanici* in 1596, which was the first to use this convention for naming of species. His criteria for classification included natural relationships, or 'affinities', which in many cases were structural.

It was in the late 1600s that plant anatomy became refined into a modern science. Italian doctor and microscopist, Marcello Malpighi, was one of the two founders of plant anatomy. In 1671 he published his *Anatomia Plantarum*, the first major advance in plant physiogamy since Aristotle. The other founder was the British doctor Nehemiah Grew. He published *An Idea of a Philosophical History of Plants* in 1672 and *The Anatomy of Plants* in 1682. Grew is credited with the recognition of plant cells, although he called them 'vesicles' and 'bladders'. He correctly identified and described the sexual organs of plants (flowers) and their parts. In the Eighteenth Century, Carl Linnaeus established taxonomy based on structure, and his early work was with plant anatomy. While the exact structural level which is to be considered to be scientifically valid for comparison and differentiation has changed with the growth of knowledge, the basic principles were established by Linnaeus. He published his master work, *Species Plantarum* in 1753.

In 1802, French botanist Charles-François Brisseau de Mirbel, published *Traité d'anatomie et de physiologie végétale* (*Treatise on Plant Anatomy and Physiology*) establishing the beginnings of the science of plant cytology.

In 1812, Johann Jacob Paul Moldenhawer published *Beyträge zur Anatomie der Pflanzen*, describing microscopic studies of plant tissues.

In 1813 a Swiss botanist, Augustin Pyrame de Candolle, published *Théorie élémentaire de la botanique*, in which he argued that plant anatomy, not physiology, ought to be the sole basis for plant classification. Using a scientific basis, he established structural criteria for defining and separating plant genera.

In 1830, Franz Meyen published *Phytotomie*, the first comprehensive review of plant anatomy.

In 1838 German botanist Matthias Jakob Schleiden, published *Contributions to Phytogenesis*, stating, "the lower plants all consist of one cell, while the higher plants are composed of (many) individual cells" thus confirming and continuing Mirbel's work.

A German-Polish botanist, Eduard Strasburger, described the mitotic process in plant cells and further demonstrated that new cell nuclei can only arise from the division of other pre-existing nuclei. His *Studien über Protoplasma* was published in 1876.

Gottlieb Haberlandt, a German botanist, studied plant physiology and classified plant tissue based upon function. On this basis, in 1884 he published *Physiologische Pflanzenanatomie* (*Physiological Plant Anatomy*) in which he described twelve types of tissue systems (absorptive, mechanical, photosynthetic, etc.).

British paleobotanists Dunkinfield Henry Scott and William Crawford Williamson described the structures of fossilized plants at the end of the Nineteenth Century. Scott's *Studies in Fossil Botany* was published in 1900.

Following Charles Darwin's *Origin of Species* a Canadian botanist, Edward Charles Jeffrey, who was studying the comparative anatomy and phylogeny of different vascular plant groups, applied the theory to plants using the form and structure of plants to establish a number of evolutionary lines. He published his *The Anatomy of Woody Plants* in 1917.

The growth of comparative plant anatomy was spearheaded by British botanist Agnes Arber. She published *Water Plants: A Study of Aquatic Angiosperms* in 1920, *Monocotyledons: A Morphological Study* in 1925, and *The Gramineae: A Study of Cereal, Bamboo and Grass* in 1934.

Following World War II, Katherine Esau published, *Plant Anatomy* (1953), which became the definitive textbook on plant structure in North American universities and elsewhere, it was still in print as of 2006. She followed up with her *Anatomy of seed plants* in 1960.

REFERENCES

1. Raven, P. H.; Evert, R. F. and Eichhorn, S. E. (2005) Biology of Plants (7th edition) W. H. Freeman, New York, page 9, ISBN 0-7167-1007-2
2. Hagemann, Wolfgang (1992). "The Relationship of Anatomy to Morphology in Plants: A New Theoretical Perspective". International Journal of Plant Sciences. 153 (3(2)): S38–S48. doi:10.1086/297062. JSTOR 2995526.
3. Evert, Ray Franklin and Esau, Katherine (2006) Esau's Plant anatomy: meristems, cells, and tissues of the plant body - their structure, function and development Wiley, Hoboken, New Jersey, page xv, ISBN 0-471-73843-3
4. See e.g. Craig, Richard & Vassilyev, Andrey. "Plant Anatomy". McGraw-Hill. Archived from the original on 24 July 2010.
5. Bolam, J. (1973). "The botanical works of Nehemiah Grew, FRS (1641-1712)". Notes and Records of the Royal Society of London. 27 (2): 219–231. doi:10.1098/rsnr.1973.0017. JSTOR 530999.

6. Thomas, Hanshaw H. (1960). "Agnes Arber, 1879–1960". Biographical Memoirs of Fellows of the Royal Society. 6: 1–11. doi:10.1098/rsbm.1960.0021. JSTOR 769330.

7. Chaffey, N. (2006). "(Book Review) Esau's Plant Anatomy, Meristems, Cells, and Tissues of the Plant Body: their Structure, Function, and Development. 3rd edn.". Annals of Botany. 99 (4): 785–786. doi:10.1093/aob/mcm015.

CHAPTER 8

Plant Ecology

Chapter Contents

- **Introduction**
- **History**
- **Distribution**
- **Biological Interactions**
- **Other Topics**

Plant Ecology and Evolution (a continuation of Belgian Journal of Botany, incorporating Systematics and Geography of Plants) is an international journal devoted to ecology, phylogenetics and systematics of all 'plant' groups in the traditional sense (including algae, cyanobacteria, fungi, myxomycetes), also covering related fields such as comparative and developmental morphology, conservation biology, evolution, phytogeography, pollen and spores, population biology, and vegetation studies.

INTRODUCTION

Plant ecology is a subdiscipline of ecology which studies the distribution and abundance of plants, the effects of environmental factors upon the abundance of plants, and the interactions among and between plants and other organisms. Examples of these are the distribution of temperate deciduous forests in North America, the effects of drought or flooding upon plant survival, and competition among desert plants for water, or effects of herds of grazing animals upon the composition of grasslands.

A global overview of the Earth's major vegetation types is provided by O.W. Archibold. He recognizes 11 major vegetation types: tropical forests, tropical savannas, arid regions (deserts), Mediterranean ecosystems, temperate forest ecosystems, temperate grasslands, coniferous forests, tundra (both polar and high mountain), terrestrial wetlands, freshwater ecosystems and coastal/marine systems. This breadth of topics shows the complexity of plant ecology, since it includes plants from floating single-celled algae up to large canopy forming trees.

A tropical plant community on Diego Garcia

One feature that defines plants is photosynthesis. One of the most important aspects of plant ecology is the role plants have played in creating the oxygenated atmosphere of earth, an event that occurred some 2 billion years ago. It can be dated by the deposition of banded iron formations, distinctive sedimentary rocks with large amounts of iron oxide. At the same time, plants began removing carbon dioxide from the atmosphere, thereby initiating the process of controlling Earth's climate. A long term trend of the Earth has been toward increasing oxygen and decreasing carbon dioxide, and many other events in the Earths history, like the first movement of life onto land, are likely tied to this sequence of events.

One of the early classic books on plant ecology was written by J.E. Weaver and F.E. Clements. It talks broadly about plant communities, and particularly the importance of forces like competition and processes like succession. Although some of the terminology is dated, this important book can still often be obtained in used book stores.

Plant ecology can also be divided by levels of organization including plant ecophysiology, plant population ecology, community ecology, ecosystem ecology, landscape ecology and biosphere ecology.

The study of plants and vegetation is complicated by their form. First, most plants are rooted in the soil, which makes it difficult to observe and measure nutrient uptake and species interactions. Second, plants often reproduce vegetatively, that is asexually, in a way that makes it difficult to distinguish individual plants. Indeed, the very concept of an individual is doubtful, since even a tree may be regarded as a large collection of linked meristems. Hence, plant ecology and animal ecology have different styles of approach to problems that involve processes like reproduction, dispersal and mutualism. Some plant ecologists have placed considerable emphasis upon trying to treat plant populations as if they were animal populations, focusing on population ecology. Many other ecologists believe that while it is useful to draw upon population ecology to solve certain scientific problems, plants demand that ecologists work with multiple perspectives, appropriate to the problem, the scale and the situation.

HISTORY

Plant ecology has its origin in the application of plant physiology to the questions raised by plant geographers.[:13–16] Carl Ludwig Willdenow was one of the first to note that similar climates produced similar types of vegetation, even when they were located in different parts of the world. Willdenow's student, Alexander von Humboldt, used physiognomy to describe vegetation types and observed that the distribution vegetation types was based on environmental factors. Later plant geographers who built upon Humboldt's work included Joakim Frederik Schouw, A.P. de Candolle, August Grisebach and Anton Kerner von Marilaun. Schouw's work, published in 1822, linked plant distributions to environmental factors (especially temperature) and established the practice of naming plant associations by adding the suffix *-etum* to the name of the dominant species. Working from herbarium collections, De Candolle searched for general rules of plant distribution and settled on using temperature as well.[:14–16] Grisebach's two-volume work, *Die Vegetation der Erde nach Ihrer Klimatischen Anordnung*, published in 1872, saw plant geography reach its "ultimate form" as a descriptive field.[:29]

Alexander von Humboldt's work connecting plant distributions with environmental factors played an important role in the genesis of the discipline of plant ecology.

Starting in the 1870s, Swiss botanist Simon Schwendener, together with his students and colleagues, established the link between plant morphology and physiological adaptations, laying the groundwork for the first ecology textbooks, Eugenius Warming's *Plantesamfund* (published in 1895) and Andreas Schimper's 1898 *Pflanzengeographie auf Physiologischer Grundlage.* Warming successfully incorporated plant morphology, physiology taxonomy and biogeography into plant geography to create the field of plant ecology. Although more morphological than physiological, Schimper's has been considered the beginning of plant physiological ecology.[:17–18] Plant ecology was initially built around static ideas of plant distribution; incorporating the concept of succession added an element to change through time to the field. Henry Chandler Cowles' studies of plant succession on the Lake Michigan sand dunes (published in 1899) and Frederic Clements' 1916 monograph on the subject established it as a key element of plant ecology.

Plant ecology developed within the wider discipline of ecology over the twentieth century. Inspired by Warming's *Plantesamfund*, Arthur Tansley set out to map British plant communities. In 1904 he teamed up with William Gardner Smith and others involved in vegetation mapping to establish the Central Committee for the Survey and Study of British Vegetation, later shortened to British Vegetation Committee. In 1913, the British Vegetation Committee organised the British Ecological Society (BES), the first professional society of ecologists. This was followed in 1917 by the establishment of the Ecological Society of America (ESA); plant ecologists formed the largest subgroup among the inaugural members of the ESA.:41

Cowles' students played an important role in the development of the field of plant ecology during the first half of the twentieth century, among them William S. Cooper, E. Lucy Braun and Edgar Transeau.:23

DISTRIBUTION

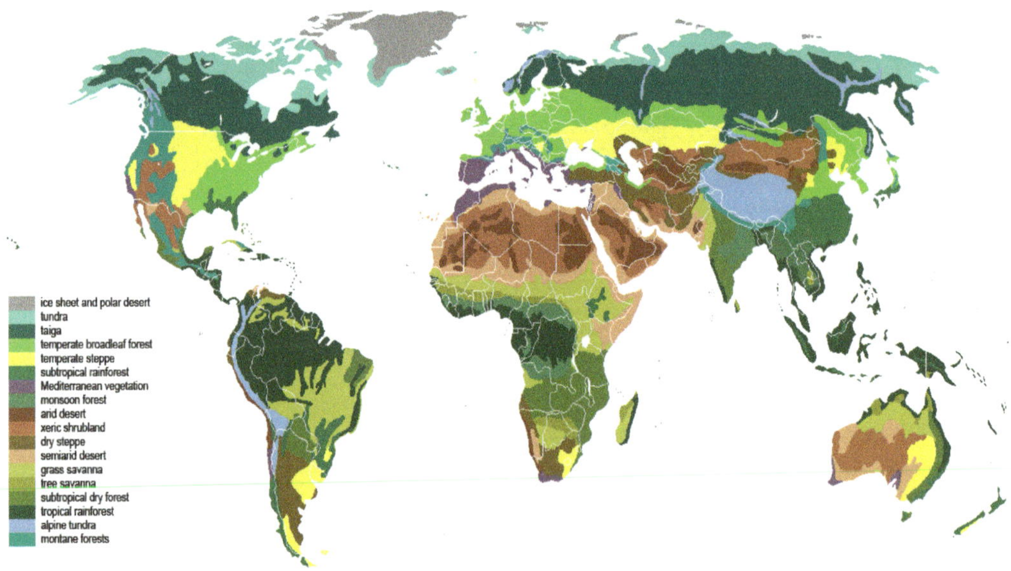

World biomes are based upon the type of dominant plant.

Plant distributions is governed by a combination of historical factors, ecophysiology and biotic interactions. The set of species that can be present at a given site is limited by historical contingency. In order to show up, a species must either have evolved in an area or dispersed there (either naturally or through human agency), and must not have gone locally extinct. The set of species present locally is further limited to those that possess the physiological adaptations to survive the environmental conditions that exist. This group is further shaped through interactions with other species.[2–3]

Plant communities are broadly distributed into biomes based on the form of the dominant plant species. For example, grasslands are dominated by grasses, while forests are dominated by trees. Biomes are determined by regional climates, mostly temperature and precipitation, and follow general latitudinal trends. Within biomes, there may be many ecological communities, which are impacted not only by climate and a variety of smaller-scale features, including soils, hydrology, and disturbance regime. Biomes also change with elevation, high elevations often resembling those found at higher latitudes.

BIOLOGICAL INTERACTIONS

Competition

Plants, like most life forms, require relatively few basic elements: carbon, hydrogen, oxygen, nitrogen, phosphorus and sulphur; hence they are known as CHNOPS life forms. There are also lesser elements needed as well, frequently termed micronutrients, such as magnesium and sodium. When plants grow in close proximity, they may deplete supplies of these elements and have a negative impact upon neighbours. Competition for resources vary from complete symmetric (all individuals receive the same amount of resources, irrespective of their size) to perfectly size symmetric (all individuals exploit the same amount of resource per unit biomass) to absolutely size-asymmetric (the largest individuals exploit all the available resource). The degree of size asymmetry has major effects on the structure and diversity of ecological communities.In many cases (perhaps most) the negative effects upon neighbours arise from size asymmetric competition for light. In other cases, there may be competition below ground for water, nitrogen, or phosphorus. To detect and measure competition, experiments are necessary; these experiments require removing neighbours, and measuring responses in the remaining plants. Many such studies are required before useful generalizations can be drawn.

Overall, it appears that light is the most important resource for which plants compete, and the increase in plant height over evolutionary time likely reflects selection for taller plants to better intercept light. Many plant communities are therefore organized into hierarchies based upon the relative competitive abilities for light. In some systems, particularly infertile or arid systems, below ground competition may be more significant. Along natural gradients of soil fertility, it is likely that the ratio of above ground to below ground competition changes, with higher above ground competition in the more fertile soils.Plants that are relatively weak competitors may escape in time (by surviving as buried seeds) or in space (by dispsersing to a new location away from strong competitors.)

In principle, it is possible to examine competition at the level of the limiting resources if a detailed knowledge of the physiological processes of the competing plants is available. However, in most terrestrial ecological studies, there is only little information on the uptake and dynamics of the resources that limit the growth of different plant species, and, instead, competition is inferred from observed negative effects of neighbouring plants without knowing precisely which resources the plants were competing for. In certain situations, plants may compete for a single growth-limiting resource, perhaps for light in agricultural systems with sufficient water and nutrients, or in dense stands of marsh vegetation, but in many natural ecosystems plants may be colimited by several resources, e.g. light, phosphorus and nitrogen at the same time.

Therefore, there are many details that remain to be uncovered, particularly the kinds of competition that arise in natural plant communities, the specific resource(s), the relative importance of different resources, and the role of other factors like stress or disturbance in regulating the importance of competition.

Mutualism

Mutualism is defined as an interaction "between two species or individuals that is beneficial to both". Probably the most widespread example in plants is the mutual beneficial relationship between plants and fungi, known as mycorrhizae. The plant is assisted with nutrient uptake, while the fungus receives carbohydrates. Some the earliest known fossil plants even have fossil mycorrhizae on their rhizomes.

The flowering plants are a group that have evolved by using two major mutualisms. First, flowers are pollinated by insects. This relationship seems to have its origins in beetles feeding on primitive flowers, eating pollen and also acting (unwittingly) as pollinators. Second, fruits are eaten by animals, and the animals then disperse the seeds. Thus, the flowering plants actually have three major types of mutualism, since most higher plants also have mycorrhizae.

Plants may also have beneficial effects upon one another, but this is less common. Examples might include "nurse plants" whose shade allows young cacti to establish. Most examples of mutualism, however, are largely beneficial to only one of the partners, and may not really be true mutualism. The term used for these more one-sided relationships, which are mostly beneficial to one participant, is facilitation. Facilitation among neighboring plants may act by reducing the negative impacts of a stressful environment. In general, facilitation is more likely to occur in physically stressful environments than in favorable environments, where competition may be the most important interaction among species.

Commensalism is similar to facilitation, in that one plant is mostly exploiting another. A familiar example is the ephiphytes which grow on branches of tropical trees, or even mosses which grow on trees in deciduous forests.

It is important to keep track of the benefits received by each species to determine the appropriate term. Although people are often fascinated by unusual examples, it is important to remember that in plants, the main mutualisms are mycorrhizae, pollination, and seed dispersal.

Herbivory

An important ecological function of plants is that they produce organic compounds for herbivores in the bottom of the food web. A large number of plant traits, from thorns to chemical defenses, can be related to the intensity of herbivory. Large herbivores can also have many effects on vegetation. These include removing selected species, creating gaps for regeneration of new individuals, recycling nutrients, and dispersing seeds. Certain ecosystem types, such as grasslands, may be dominated by the effects of large herbivores, although fire is also an equally important factor in this biome. In few cases, herbivores are capable of nearly removing all the vegetation at a site (for example, geese in the Hudson Bay Lowlands of Canada, and nutria in the marshes of Louisiana) but normally herbivores have a more selective impact, particularly when large predators control the abundance of herbivores. The usual method of studying the effects of herbivores is to build exclosures, where they cannot feed, and compare the plant communities in the exclosures to those outside over many years. Often such long term experiments show that herbivores have a significant effect upon the species that make up the plant community.

OTHER TOPICS

Abundance

The ecological success of a plant species in a specific environment may be quantified by its abundance, and depending on the life form of the plant different measures of abundance may be relevant, e.g. density, biomass, or plant cover.

The change in the abundance of a plant species may be due to both abiotic factors, e.g. climate change, or biotic factors, e.g. herbivory or interspecific competition.

Colonisation and Local Extinction

Whether a plant species is present at a local area depends on the processes of colonisation and local extinction. The probability of colonisation decreases with distance to neighboring habitats where the species is present and increases with plant abundance and fecundity in neighboring habitats and the dispersal distance of the species. The probability of local extinction decreases with abundance (both living plants and seeds in the soil seed bank).

REFERENCES

1. Keddy, Paul A. (2007). Plants and Vegetation. Cambridge: Cambridge University Press. ISBN 978-0-521-86480-0.
2. Archibold, O.W. (1995). Ecology of World Vegetation. London.: Chapman and Hall. pp. 510 p. ISBN 0-412-44290-6.
3. Weaver, J. E. and F. E. Clements. 1938. Plant Ecology. 2nd edn. New York: McGraw-Hill Book Company.
4. Schulze, Ernst-Detlef; et al. (2005). Plant Ecology. Springer. Retrieved April 24, 2012. ISBN 3-540-20833-X
5. Williams, G. C. 1975. Sex and Evolution. Monographs in Population Biology. No. 8. Princeton: Princeton University Press.

6. Harper, J. L. 1977. Population Biology of Plants. London: Academic Press.
7. van der Valk, Arnold (2011). "Origins and Development of Ecology". In Kevin deLaplante; Bryson Brown; Kent A. Peacock. Philosophy of Ecology. Handbook of the Philosophy of Science. 11. Amsterdam: Elsevier. pp. 25–48.
8. Barbour, Michael G.; Jack H. Burk; Wanna D. Pitts; Frank S. Gilliam; Mark W. Schwartz (1999). Terrestrial Plant Ecology (Third ed.). Addison Wesley Longman.
9. Cooper, W. S. (1957). "Sir Arthur Tansley and the Science of Ecology". Ecology. 38 (4): 658–659. doi:10.2307/1943136.
10. Lambers, Hans; F. Stuart Chapin III; Thijs L. Pons (2008). Plant Physiological Ecology (Second ed.).
11. Keddy, Paul A. (2001). Competition. Dordrecht: Kluwer. p. 552. ISBN 0-7923-6064-8.
12. Capser, Brenda B. and Robert. B. Jackson. 1997. Plant competition underground. Annual Review of Ecology and Systematics 28: 545–570.
13. Belcher, J., P.A. Keddy, and L. Twolan-Strutt. 1995. Root and shoot competition along a soil depth gradient. Journal of Ecology 83: 673–682
14. Twolan-Strutt, L. and P.A. Keddy. 1996. Above- and below-ground competition intensity in two contrasting wetland plant communities. Ecology 77: 259–270.
15. Craine, J. M. (2009). Resource strategies in wild plants. Princeton University Press, Princeton.
16. Grime, J. P. 1979. Plant Strategies and Vegetation Processes. Chichester: John Wiley.
17. Callaway, R. M. 1995. Positive interactions among plants (Interpreting botanical progress). The Botanical Review 61: 306–349.
18. Keddy, Paul A., Competition, 2nd ed. (2001), Kluwer, Dordrecht. 552 p.
19. Schulze, Ernst-Detlef; et al. (2005). Plant Ecology – (Section 1.10.1: Herbivory). Springer. Retrieved April 24, 2012. ISBN 3-540-20833-X
20. Keddy, P.A., Wetland Ecology: Principles and Conservation, 2nd ed. (2010), Cambridge University Press, Cambridge, UK. 497 p. Chapt. 6. Herbivory.

CHAPTER 9

Control of Plant Diseases

Chapter Contents

- **Introduction**
- **Overview**
- **Plant Pathogens**
- **Management**

Plant Pathology and Plant-Microbe Biology is a multi-faceted discipline that comprises various subjects. To reflect this diversity, the Graduate Program is organized into the following three concentrations:

INTRODUCTION

Plant pathology (also phytopathology) is the scientific study of diseases in plants caused by pathogens (infectious organisms) and environmental conditions (physiological factors). Organisms that cause infectious disease include fungi, oomycetes, bacteria, viruses, viroids, virus-like organisms, phytoplasmas, protozoa, nematodes and parasitic plants. Not included are ectoparasites like insects, mites, vertebrate, or other pests that affect plant health by consumption of plant tissues. Plant pathology also involves the study of pathogen identification, disease etiology, disease cycles, economic impact, plant disease epidemiology, plant disease resistance, how plant diseases affect humans and animals, pathosystem genetics, and management of plant diseases.

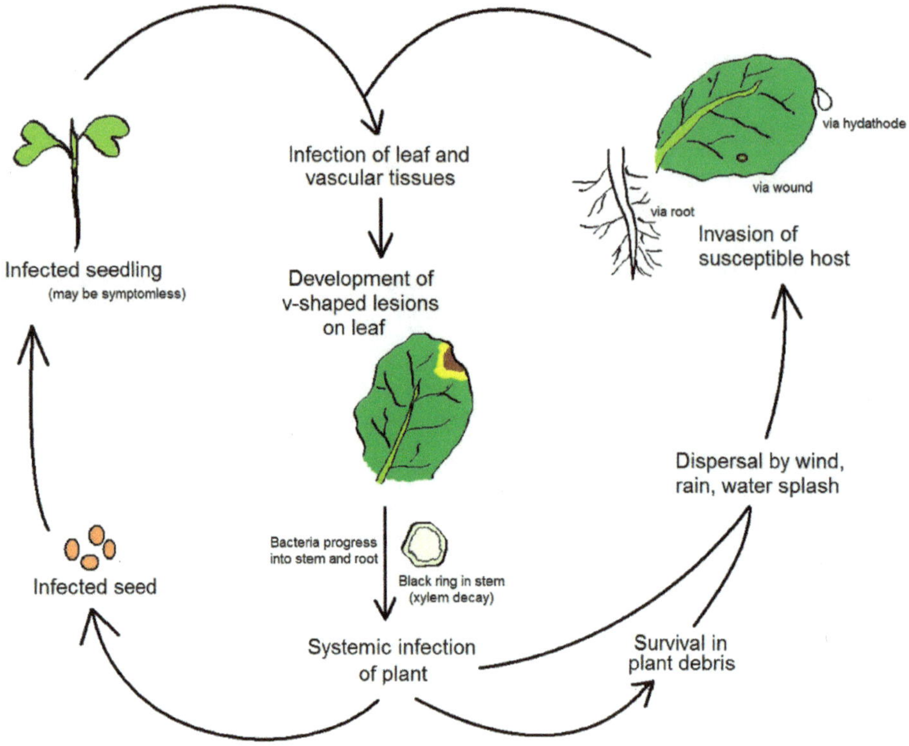

Life cycle of the black rot pathogen, *Xanthomonas campestris* pathovar *campes*

OVERVIEW

Control of plant diseases is crucial to the reliable production of food, and it provides significant reductions in agricultural use of land, water, fuel and other inputs. Plants in both natural and cultivated populations carry inherent disease resistance, but there are numerous examples of devastating plant disease impacts (see Irish potato famine, chestnut blight), as well as recurrent severe plant diseases (see rice blast, soybean cyst nematode, citrus canker). However, disease control is reasonably successful for most crops. Disease control is achieved by use of plants that have been bred for good resistance to many diseases, and by plant cultivation approaches such as crop rotation, use of pathogen-free seed, appropriate planting date and plant density, control of field moisture, and pesticide use. Across large regions and many crop species, it is estimated that diseases typically reduce plant yields by 10% every year in more developed settings, but yield loss to diseases often exceeds 20% in less developed settings. Continuing advances in the science of plant pathology are needed to improve disease control, and to keep up with changes in disease pressure caused by the ongoing evolution and movement of plant pathogens and by changes in agricultural practices. Plant diseases cause major economic losses for farmers worldwide. The Food and Agriculture Organization estimates indeed that pests and diseases are responsible for about 25% of crop loss. To solve this issue, new methods are needed to detect diseases and pests early, such as novel sensors that detect plant odours and spectroscopy and biophotonics that are able to diagnostic plant health and metabolism.

PLANT PATHOGENS

Powdery mildew, a biotrophic fungus

Fungi

Most phytopathogenic fungi belong to the Ascomycetes and the Basidiomycetes.

The fungi reproduce both sexually and asexually via the production of spores and other structures. Spores may be spread long distances by air or water, or they may be soilborne. Many soil inhabiting fungi are capable of living saprotrophically, carrying out the part of their life cycle in the soil. These are known as facultative saprotrophs.

Fungal diseases may be controlled through the use of fungicides and other agriculture practices. However, new races of fungi often evolve that are resistant to various fungicides.

Biotrophic fungal pathogens colonize living plant tissue and obtain nutrients from living host cells. Necrotrophic fungal pathogens infect and kill host tissue and extract nutrients from the dead host cells. See the powdery mildew and rice blast images, below.

Rice blast, caused by a necrotrophic fungus

Significant fungal plant pathogens include:

Ascomycetes

- *Fusarium* spp. (causal agents of Fusarium wilt disease)
- *Thielaviopsis* spp. (causal agents of: canker rot, black root rot, *Thielaviopsis* root rot)
- *Verticillium* spp.
- *Magnaporthe grisea* (causal agent of rice blast)
- *Sclerotinia sclerotiorum* (causal agent of cottony rot)

Basidiomycetes

- *Ustilago* spp. (causal agents of smut)
- *Rhizoctonia* spp.
- *Phakospora pachyrhizi* (causal agent of soybean rust)
- *Puccinia* spp. (causal agents of severe rusts of virtually all cereal grains and cultivated grasses)
- *Armillaria* spp. (the so-called honey fungus species, which are virulent pathogens of trees and produce edible mushrooms)[citation needed]

Fungus-like organisms

Oomycetes

The oomycetes are not true fungi but are fungus-like organisms. They include some of the most destructive plant pathogens including the genus *Phytophthora*, which includes the causal agents of potato late blight and sudden oak death.Particular species of oomycetes are responsible for root rot. Despite not being closely related to the fungi, the oomycetes have developed very similar infection strategies. Oomycetes are capable of using effector proteins to turn off a plant's defenses in its infection process. Plant pathologists commonly group them with fungal pathogens.

Significant oomycete plant pathogens

- *Pythium* spp.
- *Phytophthora* spp., including the causal agent of the Great Irish Famine (1845–1849)

Phytomyxea

Some slime molds in Phytomyxea cause important diseases, including club root in cabbage and its relatives and powdery scab in potatoes. These are caused by species of *Plasmodiophora* and *Spongospora*, respectively.

Bacteria

Crown gall disease caused by Agrobacterium

Most bacteria that are associated with plants are actually saprotrophic and do no harm to the plant itself. However, a small number, around 100 known species, are able to cause disease. Bacterial diseases are much more prevalent in subtropical and tropical regions of the world.

Most plant pathogenic bacteria are rod-shaped (bacilli). In order to be able to colonize the plant they have specific pathogenicity factors. Five main types of bacterial pathogenicity factors are known: uses of cell wall–degrading enzymes, toxins, effector proteins, phytohormones and exopolysaccharides.

Pathogens such as *Erwinia* species use cell wall–degrading enzymes to cause soft rot. *Agrobacterium* species change the level of auxins to cause tumours with phytohormones. Exopolysaccharides are produced by bacteria and block xylem vessels, often leading to the death of the plant.

Bacteria control the production of pathogenicity factors via quorum sensing.

Vitis vinifera with "Ca. Phytoplasma vitis" infection

Significant bacterial plant pathogens:

- Burkholderia
- Proteobacteria
- *Xanthomonas* spp.
    - *Pseudomonas* spp.

- Pseudomonas syringae pv. tomato causes tomato plants to produce less fruit, and it "continues to adapt to the tomato by minimizing its recognition by the tomato immune system."

Phytoplasmas (‹Mycoplasma-like organisms') and Spiroplasmas

Phytoplasma and *Spiroplasma* are a genre of bacteria that lack cell walls and are related to the mycoplasmas, which are human pathogens. Together they are referred to as the mollicutes. They also tend to have smaller genomes than most other bacteria. They are normally transmitted by sap-sucking insects, being transferred into the plants phloem where it reproduces.

Tobacco mosaic virus

Viruses, viroids and virus-like Organisms

There are many types of plant virus, and some are even asymptomatic. Under normal circumstances, plant viruses cause only a loss of crop yield. Therefore, it is not economically viable to try to control them, the exception being when they infect perennial species, such as fruit trees.

Most plant viruses have small, single-stranded RNA genomes. However some plant viruses also have double stranded RNA or single or double stranded DNA genomes. These genomes may encode only three or four proteins: a replicase, a coat protein, a movement protein, in order to allow cell to cell movement through plasmodesmata, and sometimes a protein that allows transmission by a vector. Plant viruses can have several more proteins and employ many different molecular translation methods.

Plant viruses are generally transmitted from plant to plant by a vector, but mechanical and seed transmission also occur. Vector transmission is often by an insect (for example, aphids), but some fungi, nematodes, and protozoa have been shown to be viral vectors. In many cases, the insect and virus are specific for virus transmission such as the beet leafhopper that transmits the curly top virus causing disease in several crop plants.

Nematodes

Root-knot nematode galls

Nematodes are small, multicellular wormlike animals. Many live freely in the soil, but there are some species that parasitize plant roots. They are a problem in tropical and subtropical regions of the world, where they may infect crops. Potato cyst nematodes (*Globodera pallida* and *G. rostochiensis*) are widely distributed in Europe and North and South America and cause $300 million worth of damage in Europe every year. Root knot nematodes have quite a large host range, they parasitize plant root systems and thus directly affect the uptake of water and nutrients needed for normal plant growth and reproduction, whereas cyst nematodes tend to be able to infect only a few species. Nematodes are able to cause radical changes in root cells in order to facilitate their lifestyle.

Protozoa and Algae

There are a few examples of plant diseases caused by protozoa (e.g., *Phytomonas*, a kinetoplastid). They are transmitted as zoospores that are very durable, and may be able to survive in a resting state in the soil for many years. They have also been shown to transmit plant viruses.

When the motile zoospores come into contact with a root hair they produce a plasmodium and invade the roots.

Some colourless parasitic algae (e.g., *Cephaleuros*) also cause plant diseases.

Parasitic Plants

Parasitic plants such as mistletoe and dodder are included in the study of phytopathology. Dodder, for example, is used as a conduit either for the transmission of viruses or virus-like agents from a host plant to a plant that is not typically a host or for an agent that is not graft-transmissible.

Common pathogenic infection methods

- **Cell wall-degrading enzymes**: These are used to break down the plant cell wall in order to release the nutrients inside.
- **Toxins**: These can be non-host-specific, which damage all plants, or host-specific, which cause damage only on a host plant.
- **Effector proteins**: These can be secreted into the extracellular environment or directly into the host cell, often via the Type three secretion system. Some effectors are known to suppress host defense processes. This can include: reducing the plants internal signaling mechanisms or reduction of phytochemicals production. Bacteria, fungus and oomycetes are known for this function.

Physiological Plant Disorders

Significant abiotic disorders can be caused by:

Natural

Drought

Frost damage and breakage by snow and hail

Flooding and poor drainage

Nutrient deficiency

Salt deposition and other soluble mineral excesses (e.g., gypsum)

Wind (windburn and breakage by hurricanes and tornadoes)

Lightning and wildfire (also often man-made)

Man-made (arguably not abiotic, but usually regarded as such)

Soil compaction

Pollution of air, soil, or both

Salt from winter road salt application or irrigation

Herbicide over-application

Poor education and training of people working with plants (e.g. lawnmower damage to trees)

Vandalism

Orchid leaves with viral infections

MANAGEMENT

Quarantine

A diseased patch of vegetation or individual plants can be isolated from other, healthy growth. Specimens may be destroyed or relocated into a greenhouse for treatment or study. Another option is to avoid the introduction of harmful nonnative organisms by controlling all human traffic and activity (e.g., AQIS), although legislation and enforcement are crucial in order to ensure lasting effectiveness.

Cultural

Farming in some societies is kept on a small scale, tended by peoples whose culture includes farming traditions going back to ancient times. (An example of such traditions would be lifelong training in techniques of plot terracing, weather anticipation and response, fertilization, grafting, seed care, and dedicated gardening.) Plants that are intently monitored often benefit from not only active external protection but also a greater overall vigor. While primitive in the sense of being the most labor-intensive solution by far, where practical or necessary it is more than adequate.

Plant resistance

Sophisticated agricultural developments now allow growers to choose from among systematically cross-bred species to ensure the greatest hardiness in their crops, as suited for a particular region's pathological profile. Breeding practices have been perfected over centuries, but with the advent of genetic manipulation even finer control of a crop's immunity traits is possible. The engineering of food plants may be less rewarding, however, as higher output is frequently offset by popular suspicion and negative opinion about this "tampering" with nature.

Chemical

(See: pesticide application) Many natural and synthetic compounds can be employed to combat the above threats. This method works by directly eliminating disease-causing organisms or curbing their spread; however, it has been shown to have too broad an effect, typically, to be good for the local ecosystem. From an economic standpoint, all but the simplest natural additives may disqualify a product from "organic" status, potentially reducing the value of the yield.

Biological

Crop rotation may be an effective means to prevent a parasitic population from becoming well-established, as an organism affecting leaves would be starved when the leafy crop is replaced by a tuberous type, etc. Other means to undermine parasites without attacking them directly may exist.

Integrated

The use of two or more of these methods in combination offers a higher chance of effectiveness.

REFERENCES

1. Agrios, George N. (1972). Plant Pathology (3rd ed.). Academic Press.
2. Martinelli, F., Scalenghe, R., Davino, S., Panno, S., Scuderi, G., Ruisi, P., Villa, P., Stroppiana, D., Boschetti, M., Goulart, L.R., Davis, C.E., Dandekar, A.M. (2014). "Advanced methods of plant disease detection. A review". Agronomy for Sustainable Development. 35: 1–25. doi:10.1007/s13593-014-0246-1.
3. Nicole Davis (September 9, 2009). "Genome of Irish potato famine pathogen decoded". Haas et al. Broad Institute of MIT and Harvard. Retrieved 24 July 2012.
4. Kamoun, S.; Furzer, O.; Jones, J. D. G.; Judelson, H. S.; Ali, G. S.; Dalio, R. J. D.; Roy, S. G.; Schena, L.; Zambounis, A.; Panabières, F.; Cahill, D.; Ruocco, M.; Figueiredo, A.; Chen, X. R.; Hulvey, J.; Stam, R.; Lamour, K.; Gijzen, M.; Tyler, B. M.; Grünwald, N. J.; Mukhtar, M. S.; Tomé, D. F. A.; Tör, M.; Van Den Ackerveken, G.; McDowell, J.; Daayf, F.; Fry, W. E.; Lindqvist-Kreuze, H.; Meijer, H. J. G.; et al. (2015). "The Top 10 oomycete pathogens in molecular plant pathology". Molecular Plant Pathology: n/a. doi:10.1111/mpp.12190.
5. Grünwald, N. J.; Goss, E. M.; Press, C. M. (2008). "Phytophthora ramorum: A pathogen with a remarkably wide host range causing sudden oak death on oaks and ramorum blight on woody ornamentals". Molecular Plant Pathology. 9 (6): 729. doi:10.1111/J.1364-3703.2008.00500.X.
6. "Scientists discover how deadly fungal microbes enter host cells". (VBI) at Virginia Tech affiliates. Physorg. July 22, 2010. Retrieved July 31, 2012.
7. Jackson RW (editor). (2009). Plant Pathogenic Bacteria: Genomics and Molecular Biology. Caister Academic Press. ISBN 978-1-904455-37-0.
8. Burkholder (October 1948). "Bacteria as Plant Pathogens". Annual Review of Microbiology. Cornell University. 2: 389–412. doi:10.1146/annurev.mi.02.100148.002133. PMID 18104350.
9. "Research team unravels tomato pathogen's tricks of the trade". Virginia Tech. 2011.
10. Creamer, Rebecca; H. Hubble; A. Lewis (May 2005). "Curtovirus Infection of Chile Pepper in New Mexico". Plant Diseases. 89 (5): 480–486. doi:10.1094/PD-89-0480.
11. Huynh, Bao-Lam; Matthews, William C.; Ehlers, Jeffrey D.; Lucas, Mitchell R.; Santos, Jansen R. P.; Ndeve, Arsenio; Close, Timothy J.; Roberts, Philip A. (2015-10-08). "A major QTL corresponding to the Rk locus for resistance to root-knot nematodes in cowpea (Vigna unguiculata L. Walp.)". Theoretical and Applied Genetics. 129 (1): 87–95. doi:10.1007/s00122-015-2611-0. ISSN

0040-5752. PMC 4703619Freely accessible. PMID 26450274.

12. Jankevicius, J.V. et al. Ciclo biológico de Phytomonas / Biological cycle of Phytomonas. Memórias do Instituto Oswaldo Cruz, Rio de Janeiro, v. 83, supl. 1, 1988.
13. Ma, Winbo (March 28, 2011). "How do plants fight disease? Breakthrough research by UC Riverside plant pathologist offers a clue". UC Riverside.
14. "1st large-scale map of a plant's protein network addresses evolution, disease process". Dana-Farber Cancer Institute. July 29, 2011. Retrieved 24 July 2012.
15. "History of Plant Pathology". Retrieved 5 February 2015.
16. "Plasmopara viticola, the Cause of Downy Mildew of Grapes". The Origin of Plant Pathology and The Potato Famine, and Other Stories of Plant Diseases. Retrieved 4 February 2015.
17. "Fusarium oxysporum : The End of the Banana Industry?". The Origin of Plant Pathology and The Potato Famine, and Other Stories of Plant Diseases. Retrieved 4 February 2015.
18. "Plant Diseases - History Of Plant Pathology". Retrieved 5 February 2015.
19. Aisnworth GC. (1981). Introduction to the History of Plant Pathology. Cambridge University Press. ISBN 0-521-23032-2.

Index